SPECTRUMS

for "amplitude modulation," which means the signal—the music, or the sportscaster yammering away—is encoded by increasing and decreasing the volume or intensity very slightly but very quickly. As we learned earlier, a light's intensity (its strength, or amplitude) is based on how many photons are being transmitted. Your radio senses those tiny changes and converts them into (you hope) a pleasant sound.

Switch to 103.7 on the FM dial, and your radio begins to pick up light waves at 103.7 MHz, or 103,700,000 cycles per second. Here you experience a different kind of signal: "frequency modulation," in which the frequency of the wave is altered up and down while maintaining its intensity. (More precisely, one wave with a varying frequency is merged with another, steady wave, and the result—an extremely complex wave—is transmitted, received, and pulled apart again.)

Awash in Light Radio broadcasts, satellite television transmissions, cell phone conversations, GPS signals, police and fire alerts, WiFi computer networks, radio-controlled toys, airport radar, garage door openers—we are constantly bombarded by light of varying wavelengths, energies, frequencies. And yet, strangely, most of this human-made radiation passes right through our bodies, even through the walls of our buildings, without affecting us or even slowing down. Why?

Light—which, remember, is made of photons, those tiny packages of energy that behave like both waves and particles—is unique in the universe in its ability to be both exceedingly tiny and extremely large. This range of size largely explains its ability to travel. Because each wave from an AM radio station is longer than a football field, the relatively small and sparse molecules that make up a wall don't have enough presence to impact it much. But line the wall with thick metal—a material with dense connections of electrically bound atoms—and those long waves can't get through.

You see, matter (like an atom or a molecule) can interact with a light wave in one of three ways: let it pass by, absorb it, or bend it. Which of these occurs is based on the wavelength of the light versus the type, size, density, and structure of the material. An X-ray has a tiny wavelength, so it passes through our skin like a bike zooming through a forest of widely spaced trees. Every so often it might hit a branch and cause a little damage, but in general it won't stop until it reaches a thick bramble, like the molecules in your bones. At that point, the light will likely be either bent (forcing a quick change of course) or absorbed (like a bike crash).

Microwaves, as we saw earlier, are far longer than X-rays—small enough to be easily absorbed by some food molecules but large enough that they cannot escape from the oven into your kitchen.

Visible light, on the other hand, consists of wavelengths that are just the right size to be absorbed or reflected by most matter around us. This is, as scientists like to say, evolutionarily advantageous: If our eyes were tuned to see radio waves instead, we'd be constantly banging into "invisible" things around us. But because we see the visible light spectrum, a ripe banana tends to absorb the "blue" wavelengths and reflect "red" and "green" waves into our eyes, causing us to see a yellow object.

Rocks appear solid because they, too, reflect and absorb various frequencies of light. But if you grind rocks to sand and melt the sand to make glass, you've changed the molecular structure so radically that wavelengths of visible light can now pass through it largely unimpeded. Yet those same molecules that let visible light pass through are also absorbing the slightly shorter wavelengths of ultraviolet light. So the size of the wave matters, but not as much as the material composition.

Now You See It . . . Light is such an inherent part of our everyday experience—whether through the colors we see or the heat we feel—that it's easy to forget how fundamental it is to the underlying structure of our universe, and how little we truly understand it. After all, electromagnetism is not just the study of magnets and generators and light waves. It is considered one of the four fundamental forces of nature, along with gravity and the strong and weak nuclear interactions. These are the most basic physical forces that hold our universe together and that cannot be described by any other, further reduced explanation.

Electromagnetic force pulls atoms into molecules and holds them together across enormous (to an atom) distances, forming what (from our size) appear to be solid objects. Without that subtle attraction, your chair, your floor, you, Earth would come apart as nothing but gas. At the core of this electromagnetic force is the lowly photon—called a quanta by Einstein as he laid the groundwork for quantum mechanics. Photons—known as the carrier particles of electromagnetism—are literally what make our universe possible.

Light—the oscillation of electromagnetic waves, the transmission of photons—is like the lifeblood of the cosmos, carrying packets of energy from one atom to another, or from one galaxy to another, through billions of years of space. Although it appears from our perspective that there is no medium in the vacuum between the stars (or between the atoms) through which these waves could be held and passed, that misses the point: Space itself is the medium. We are the medium. And we are the receivers.

SOUND

These go to eleven.

Nigel Tufnel, *This Is Spinal Tap*

SOUND IS MOVEMENT. SPECIFICALLY, SOUND IS THE MOVEMENT of molecules in a medium. And our sense of hearing is actually an extension of feeling—as though our ears were fingers that could reach out and stroke the ripples that run through the air around us. Hearing sound, like perceiving light, is another one of our evolutionarily clever ways of stretching our awareness beyond our boundaries, whether in search of food, concern for danger, or desire for connection.

If you stand next to a speaker playing loud music, you can often feel the sounds against your skin—rumbling low bass notes and buzzing high-pitched tones. You're feeling waves of air pressure, molecules moving and bouncing into each other. Our ears are so sensitive to these changes in pressure that we can detect when air molecules move one-tenth their diameter—that's about one-millionth the size of the smallest dust speck you can see.

The molecules move because of energy. Somewhere, somehow, something releases energy—a string is plucked, a dog barks, a reed flutters, or what have you. This is physical, kinetic energy; it moves, and the movement flows outward in waves. One molecule pushes into another, but the molecules can't move far until they affect another molecule, like people crowding to get out of a theater after the show.

So as one molecule pushes the next, each moves only a little, but the energy continues. It propagates. It travels. It transmits, like a

> **"I don't care much about music. What I like is sounds."**
>
> —Dizzy Gillespie, jazz musician

message being passed from one person to another. Each passing wears down the energy a little, but with enough thrust, enough power, it can travel a long way—perhaps even all the way to your ear.

The Medium Is the Message Of course, if there is no medium, there is no sound. The great Greek thinker Aristotle, student of Plato and teacher of Alexander the Great, first pointed out that we hear sound through the medium of air. Want the ultimate in soundproofing for your office or apartment? Encase yourself within a perfect vacuum of space, for where there is nothing (literally no-thing, no molecules to push around), there is no sound.

The statement in the movie *Alien* that "in space, no one can hear you scream" is true; those movie scenes in which inhabitants of one spaceship can hear a sound in another nearby spaceship are impossible. Without some medium—some solid, gas, or liquid to transmit the energy waves from one location to another—there is no message.

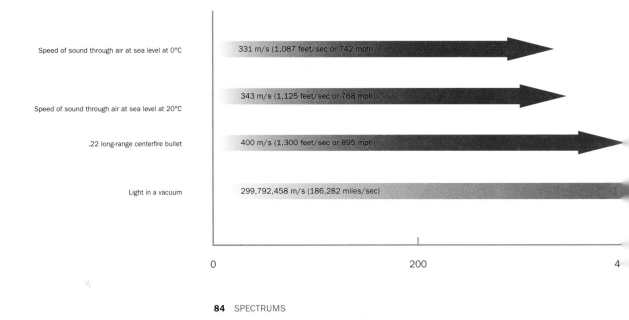

Speed of sound through air at sea level at 0°C — 331 m/s (1,087 feet/sec or 742 mph)

Speed of sound through air at sea level at 20°C — 343 m/s (1,125 feet/sec or 768 mph)

.22 long-range centerfire bullet — 400 m/s (1,300 feet/sec or 895 mph)

Light in a vacuum — 299,792,458 m/s (186,282 miles/sec)

0   200   4

This is distinctly different from electromagnetic energy. Electromagnetic waves—whether X-rays, radio waves, or even the small sliver of the spectrum called visible light—don't require molecules and can voyage through space indefinitely, easily traveling the 24 trillion miles (40 trillion km) of nothingness between us and our nearest star, Proxima Centauri. But a sound, no matter how great, peters out at the edge of the atmosphere.

Sound isn't limited to earth's atmosphere, of course. We are awash in the radiation of our sun, but we rarely consider the extraordinary noise that this ball of gas is making. You think a crackling fire is loud on a rainy night? The turbulence of gas and plasma churning, burning, exploding, at 16,000,000°C is staggering, sending incomprehensible shockwaves outward in all directions through the sun's mixture of hydrogen, helium, oxygen, and gaseous metals. The sound energy pounds out toward the surface of the burning ball, and then, as it reaches the boundary of the sun's atmosphere, the gas eventually stops, and with it the sound. Between our beneficent star and earth is silence.

600   800   1,000

Of course, if you were actually close enough to hear the tolling of the sun, you would be vaporized. But just as you can "see" sound by watching a rattling window, scientists can see the sounds of the sun from 93 million miles away using precise Doppler measurements taken by the Solar and Heliospheric Observatory (SOHO) satellite. The surface of the sun vibrates with a complex set of resonances. Although the vibrations are too low for the human ear to hear, they can be sped up—compressing forty days worth of sound into a few seconds. The result is like an eerie bell, or a Buddhist bowl gong slowly, endlessly ringing into space.

Sound travels through water at different speeds, depending on temperature. So scientists can use hydrophones (underwater microphones) to determine water temperature by measuring the speed of sound in a particular location.

The Speed of Sound Because sound relies on molecules bouncing into one another, passing along their energy wave, it takes time for it to travel from one place to another. Sound moves fast, but not nearly as quickly as light—in fact, not even as fast as a bullet from a high-powered rifle.

It's relatively easy to measure the speed of sound: Place two people a mile apart, giving one a sports starter pistol with blanks and the other person binoculars and a stopwatch. When the pistol is fired (watch for smoke), start the stopwatch, then stop it when you hear the sound. Intuition says we probably couldn't start and stop it quickly enough, but you'll actually count almost five seconds before hearing the blast.

This exercise explains a common game during thunderstorms: Start counting seconds when you see a flash of lightning and stop when you hear the thunder. Divide the total number of seconds by 5 to find out how many miles away the lightning struck. (It's not 1 mile per second, as some people mistakenly believe.) Or divide by 3 to find the number of kilometers.

With careful measurements, you'll find that sound travels through air about 1,125 feet (343 m) in a second. That works out to about 67,500 feet (20,580 m) per minute, or 767 miles per hour (1,234 km/h).

Notice the word *about*. The speed of sound can change significantly depending on factors such as temperature. On a freezing cold day, it drops to 330 m/sec. On a hot day it can increase to 350 m/sec. The difference is due to the rate the molecules are moving through the gas we call air. As the sun warms the air, the molecules move faster, bounce into each other more often, and allow signals (such as sound wave energy) to transmit faster.

We need to be clear here: The molecules are not traveling very far. When a book falls off a table across the room, the affected molecules don't travel from the book to your ear. That would be wind, not sound. But like a series of billiard balls ricocheting across a pool table, the energy within the sound finds its way to you.

The makeup of the air also affects the speed of sound. For example, the speed of sound is faster in helium gas, which contains far lighter molecules than air, leading to the time-honored party game of talking with a lungful of helium. The pitch of your voice actually stays the same, but it moves through the gas faster, causing it to sound sped-up, like Donald Duck. If you're lucky enough to have a supply of heavier-than-air xenon gas handy, you can reverse the effect and sound like a "slow-talking cowboy" through the magic of slowing the speed of sound.

Sound can travel through liquids and solids, too, and at a very different speed than gas. The exact speed changes radically, based partly on the elasticity and density of the material. In freshwater, where the molecules are packed together in a slightly sticky and viscous solution, sound waves travel more than four times faster than in air: 1,482 meters per second (about 3,315 mph). In seawater, the speed of sound increases by a few percent, depending on temperature, depth, and salinity.

In a solid, where molecules are bound even more securely and one can barely move without affecting its neighbor, energy can pass even faster. A soft, relatively elastic material such as lead transmits sound waves at just over 2,000 m/sec, but in steel the speed of sound is about 5,960 m/s (that's 21,450 km/h or 13,300 mph). That's why you

Carbon dioxide	259
Air, 0°C	331
Nitrogen	334
Air, 20°C	343
Cork	500
Helium	965
Ethyl alcohol	1207
Water, distilled	1497
Water, sea	1531
Soft tissues	1540
Rubber	1600
Lead	2160
Gold	3240
Brick	3650
Marble	3810
Wood, oak	3850
Wood, maple	4110
Copper	4760
Glass, pyrex	5640
Steel, stainless	5790
Granite	5950
Aluminum	6420
Beryllium	12,890

▲ **Speed of sound in different materials (m/sec)**

can hear a train approach by pressing your ear to the tracks; you'll hear it 17 times faster through metal than through the air. If the tracks were made of an incredibly hard substance, such as beryllium or diamond, you'd hear it almost 40 times faster than through air.

Curiously, sound does not tend to pass very well from one medium to another. Sound waves act like light waves in this respect. Just as light reflects and refracts when moving from one medium (like air) into another (say, water), sound waves bend and bounce at these boundaries. So a crack of a hammer against the end of a long steel bar may travel beautifully through the metal, but little of it will transfer into the air on the other side. Similarly, sound waves from the air don't penetrate well into water, and vice versa. Anglers, take note: Feel free to chat with your buddies; the sound won't scare the fish!

A noisy object creates sound waves that expand out in all directions, like tiny ripples in a still pond. If the object starts to move, then those ripples keep expanding, but they become elongated, falling farther behind the object than in front of it, similar to the wake behind a boat. Just as waves don't hit the shore until long after the boat has passed, we don't hear the rumble of a jet airliner flying far overhead until the plane is almost out of sight.

But a funny thing happens as a jet plane nears the speed of sound: The sound waves compress and bunch up like wrinkled cloth in front of the nose of the aircraft. It's as though the molecules of air can't get out of the way quickly enough, and the pilot begins to experience increased aerodynamic drag, like an invisible hand pushing back.

When fighter jets first encountered this during World War Two, some erroneously believed that supersonic flight—flying beyond the speed of sound—was impossible, like traveling faster than the speed of light. However, the soldiers need only to have looked at their artillery. Bullets easily pass the speed of sound, flying through the air at speeds as fast as 5,000 feet/second (1,500 m/sec).

The exploding charge that propels ammunition creates a sound, but the bang actually stems from something unexpected: The

> "Wherever we are, what we hear is mostly noise. When we ignore it, it disturbs us. When we listen to it, we find it fascinating."
>
> —John Cage, composer

▲ Lockheed SR-71 Blackbird

pressure waves that build up in front of a supersonic object can no longer get away from their source. The result of this compression is an intense shock wave that travels through the air, often referred to as a sonic boom. Even a gun with a muzzle silencer cannot stop the sound of the bullet as it pierces the air, generating an intensely sharp peak in air pressure, like a thin but powerful wall of sound.

Modern fighter jets create similar, though obviously much larger, sonic booms as they move at transonic speeds. Though you may hear a sudden rumble that hits and then passes, the shock wave actually continues, following behind the plane like a sharp-edged sonic shadow that extends until the air pressure can sufficiently dissipate.

When dealing with very fast-moving objects, it's often helpful to discuss their speed as multiples of Mach, named in honor of the

A bullwhip snapped properly creates a tiny sonic boom as the end of the whip (the "cracker") quickly flips around faster than the speed of sound.

▲ Heinrich Hertz

physicist and philosopher Ernst Mach (1838–1916), who studied (among many other things) sound and ballistics. One half Mach is half the speed of sound, Mach 2 is twice the speed of sound, and so on. For many years, the fastest airplane was the Lockheed SR-71 Blackbird, which broke the speed record at Mach 3 in 1964. Forty years later, NASA's X-43A scramjet busted the doors off the record, with speeds up to Mach 9.6—nearly 7,000 mph.*

In order to escape Earth's orbit, a rocket needs to go even faster—about Mach 23 (17,000 mph), or twice that to propel itself to the moon. Obviously, the sound generated by these engines creates a sonic boom that is very, very loud.

How Loud Is Loud? Why are some sounds louder than others? Like ocean waves, sound waves are each created with a crest and a trough, and their sizes—the difference between their peaks and the surrounding air pressure (or sea level, using that analogy)—is called the wave's amplitude. For example, musicians know you take a small sound signal and make it huge with an amp, or amplifier—a device that increases the amplitude of the wave.

The bigger the difference in air pressure, the bigger the wave, the louder the sound.

We're not talking about huge differences in pressure here. Pressure is often measured in pascals (Pa), and we live in a bubble of air pressurized at 101,325 Pa. Let's say a sound wave is moving toward us. The air pressure momentarily increases and decreases by a tiny amount (remember, there is always a crest and a trough, technically called compression and rarefaction). If the pressure changes by 2 Pa (just 0.002 percent), we hear it—not as a whisper, as you might expect, but as the deafening sound of a jackhammer breaking through stone. A quiet conversation alters air pressure by as little as 0.0005 Pa (less than 5 ten-millionths of 1 percent).

*Though of course the X-43A and other recent hypersonic aircraft have all been unmanned, so the Blackbird could still technically be considered the fastest airplane.

Intensity of Selected Sounds

Loudness/Intensity (dB)	Source
<0	Silence
0–10	Faintest noise humans can hear
10–20	Normal breathing in quiet room, rustling leaves
20–30	Whispering at 5 feet
30–40	Library or calm room
40–50	Quiet office, normal talking or residential area
50–60	Dishwasher, electric toothbrush, rainfall, sewing machine
60–70	Air conditioner, automobile interior, background music, normal conversation, TV, vacuum cleaner
70–80	Coffee grinder, freeway traffic, garbage disposal, hair dryer
80–90*	Blender, doorbell, food processor, lawn mower, machine tools, noisy restaurant, whistling kettle
90–100	Shouted conversation, tractor, truck
100–110	Boom box, factory machinery, motorcycle, school dance, snowblower, snowmobile, subway train
110–120	Ambulance siren, car horn, chain saw, disco, jet plane on ramp, rock concert, shouting in ear
120–130	Heavy machinery, pneumatic drills, stock car races, thunder (short term hearing damage)
130–140	Air raid siren, jackhammer (threshold of pain)
140–150	Jet airplane taking off
150–160	Artillery fire at 500 feet
160–170	Fireworks, handgun, rifle
170–180	Shotgun, stun grenade
180–190	Rocket launch, volcanic eruption
194	Theoretical limit for undistorted sound in air

*Employers in the United States must provide hearing protectors to all workers exposed to continuous noise levels of 85 dB or above.

Perhaps you can begin to see how sensitive our ears are. Due to a complex system of interrelated bones and membranes in our middle and inner ear, we can pick up a change of less than a billionth of the atmospheric pressure, where air molecules are moved less than the diameter of an atom.

Instead of talking in pascals, most people refer to loudness using a different value: the decibel (dB), measuring one-tenth of a "bel" (named in honor of the telecommunications pioneer Alexander Graham Bell). The decibel is an almost entirely humancentric measurement: Zero decibels marks not some universal constant but the lower limit of human hearing—the faintest sound we can detect. Below that, you can't tell the difference between sound and air molecules just randomly bumping up against the eardrum.

The logarithmic nature of the decibel system means that for each additional 10 decibels, *ten times* more power is required, but it *doubles* the perceived loudness of a sound. That is, a normal conversation (about 40 dB) is about twice as loud as a quiet library (about 30 dB), but that 30 decibels reflect 1,000 times more power than near silence. A large truck driving by can throw 94 decibels—carrying almost ten million (10^7) times the power of a whisper.

In a famous 1976 concert, The Who was measured at 126 dB, 100 feet (30 m) from the stage. More recently, the band KISS hit 136 dB—the equivalent of standing next to a jet airplane taking off—during a 2009 Canadian concert, just before being forced to "turn it down" by local law enforcement. That's 17,000 times louder and 10 trillion times more powerful than a heartbeat. One can only hope earplugs were liberally distributed before these shows, as permanent hearing damage can be caused by sound above 120 decibels.

Of course, a rock concert is nothing compared with the new international "sport" of dB drag racing—where competitors build cars that contain virtually nothing but an engine and audio equipment. The goal is to create the loudest car, if only for a few seconds. The vehicles have two-inch-thick windows and doors bolted and clamped closed so as not to rattle off their hinges. Participants stand outside and throw a switch to create a pulse of noise so powerful it can literally melt the metal in the speakers. The world-record car, at about 180 dB, is 60 times louder and reflects a million times more power than a typical concert.

In fact, that car nearly ties the loudest sound on record, which most historians identify as the 1883 volcanic eruption at Krakatau, Indonesia. The cataclysm, in which most of the island was destroyed and ash was propelled 50 miles (80 km) high, had an intensity of just over 180 dB and was audible 3,000 miles (5,000 km) away in Mauritius. The shock wave traveled even farther, reverberating literally around the world over the next five days.

But could a sound be even louder than that? It depends on how you define *sound*. There is no maximum strength of a shock wave. Blow up a few hundred pounds of TNT, and you'll create about 200 dB of pressure—a wave of such power that it would likely kill any human nearby. Nuclear explosions reach over 275 dB. However, if you limit the definition of sound to a pressure wave with a crest and a trough, a signal that conveys a message through the air beyond a distorted, deadly boom, then you're limited to a wave no bigger than atmospheric pressure itself. That is, the rarefaction (the low point on the wave) cannot drop below zero Pa, the pressure in a vacuum. And a drop from normal atmospheric pressure of 101,325 Pa to zero Pa results in a maximum possible sound volume of 194 dB.

Sound power is measured in watts per square meter or W/m². The ratio of power required to generate the faintest sound we can hear, up to the level of "Ow, that hurts!" is 1:100,000,000,000,000 (a hundred million million). This is yet another example of how incredibly dynamic and sensitive human hearing is. **The farther from a sound** source, the less loud you hear it. Specifically, the sound intensity drops by about 6 dB each time you double the distance.

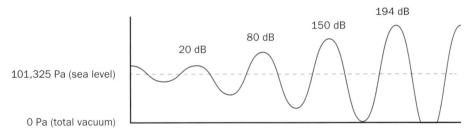

▲ As volume increases, sound waves eventually get clipped.

What's the Frequency, Kenneth? Our ears are clearly sensitive to a sound wave's amplitude, but they're just as sensitive to the distance from the crest of one wave to the crest of the next—the wavelength. This measurement, along with the speed the sound is traveling, determines the amount of time it takes for one full wave to crash into our eardrums, like waiting for one ocean wave to finish before the next comes ashore.

That math-loving Greek Pythagoras, about 2,500 years ago, first observed that a string held taut and plucked vibrates at a particular rate. The vibration of the string then causes a sound. But tighten the string, use a thinner filament, or shorten it, and the pitch rises to a higher note. A string half as long produces a sound exactly one octave higher. A string twice as long is an octave lower.

A string is a helpful tool to understand waves because we can literally see it quiver. A loose, flapping wire makes no sound we can hear, but tighten it until it's vibrating more than 20 times per second, and you perceive a moan; keep tightening, and the sound rises to a groan, then a growl, each slightly higher in pitch. The tone is based entirely on the frequency of the waves—that is, the number of vibrations each second. Lower frequencies—longer waves and fewer cycles per second—sound lower to us. Higher frequencies (shorter waves, more coming at us each second) sound high-pitched.

We see examples of vibrations all around us. A bird flapping its wings 2 or 3 times per second creates no sound we can hear. But a bumblebee wing, flapping about 200 times per second, creates a low hum. A mosquito wing, moving 600 times per second, is an annoying whine. Again, our hearing comes to our rescue, enabling us to "reach out" and find the insect invader.

Scientists replace the phrase "waves or cycles per second" with the simple word *hertz* (Hz). You could say a clock ticks off time at 1 hertz (once per second), though the tick or tock it makes is actually a sound wave vibrating far faster. In fact, the lowest pitch humans can hear is a wave vibrating around 15 Hz. A tone at 30 Hz sounds about the same, but an octave higher; the same can be said for 60, about

Sound Frequency

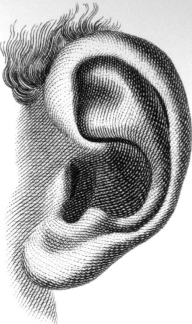

Frequency	Sound phenomenon
0.1–2 Thz	SASER (sound laser, in development)
1–20 MHz	Medical ultrasound
25–100 kHz	Bat sonar clicks
40–50 kHz	Ultrasonic cleaning
32.768 kHz	Quartz timing crystal
18–20 kHz	Upper limit of human hearing
4–5 kHz	Field cricket (Teleogryllus oceanicus)
2048 Hz	C7 scientific scale, highest note of a soprano singer (approximate)
440 Hz	A4 American standard pitch, TV test pattern tone
435 Hz	A4 international pitch
261.63 Hz	"Middle C" (C4 in the American standard pitch)
256 Hz	C4 scientific scale, typical fundamental frequency for female vocal cords
128 Hz	C3 scientific scale, typical fundamental frequency for male vocal cords
64 Hz	C2 scientific scale, lowest note of a bass singer (approximate)
50 Hz	Ruby-throated hummingbird in flight
20–50 Hz	Cat purr
20 Hz	Lower limit of human hearing
17–30 Hz	Blue and fin whales are the loudest marine sound in this range
1–5 Hz	Tornadoes

the throb of a hummingbird flying by. Double that is 120, about the typical frequency of a man's voice. Another doubling brings us to the range of a woman's voice, though the human voice can actually span from about 80 to 1,100 Hz.

When the sound waves are compressed into the thousands-per-second range, we start measuring in kilohertz (kHz). Children can easily hear sound up to 20 kHz (20,000 wave cycles per second), though that ability tends to wear out for one reason or another until, at middle age, we tend not to be able to hear anything higher than 15 or 16 kHz. Marketers have taken advantage of the difference. A Welsh security company created the Mosquito sound repellent that emits screeches in the 17 kHz range, designed to repel teenagers from loitering in front of shops. Of course, the tables were soon turned when the high-frequency sound was converted into a cell phone ringtone—one that kids can hear but adults (such as teachers and parents) cannot.

Note that this range, from 20 waves per second to 20,000, is an extraordinary span, reaching over ten octaves (where each octave

Weather conditions can also make sound travel farther or shorter distances—but not for the reason you think. A strong wind won't carry a sound wave along faster or push it backward like it would a material object. After all, the speed of sound is far faster than the speed of the wind! However, wind shear—when the direction that air is traveling changes suddenly from one part of the atmosphere to another—can affect how sound travels. Sound typically bends upward as it travels out, but with an appropriate wind shear, it will reflect downward again, like a stone skipping on a pond—causing someone farther away to be able to hear it. Similarly, temperature inversions— where a layer of warm air rests on top of a bubble of cooler air—can also cause loud sounds to travel much farther than expected. An explosion at an English oil depot could be heard a surprising 200 mi (320 km) away in the Netherlands as the sound ricocheted off a layer of air. People who were much closer to the blast did not hear it, as the sound traveled over their heads, creating a "sound shadow."

represents a doubling of frequency). Compare that with our eyes, sensitive to only a single octave of the electromagnetic spectrum between about 400 and 780 terahertz.

Plus, imagine the speed at which our ears are processing information. First, the sound waves are captured by our pinna (those are the fleshy things sticking out from the side of your head), which act as sophisticated sound-processing gear, cleverly amplifying and filtering sounds before focusing them into the ear canal. The waves then vibrate a thin but rigid piece of skin, technically called the tympanic membrane but commonly called the eardrum.

While the analogy of a drum seems apt at first, with the sound waves beating against it, the truth is that the compression and rarefaction (the positive and negative changes in pressure) actually push and pull at the eardrum. This physical movement is then transferred to an astonishingly complex mechanism in which the three tiniest bones in your body (you remember from school: hammer, anvil, stirrup) act like a hydraulic lever, amplifying the faint sound signals 22 times while pressing against the fluid-filled snail-shell-shaped cochlea. Finally, the waves pass through the cochlear fluid to stimulate more than 20,000 minuscule hairs, like underwater currents wagging long strands of seaweed attached to the ocean floor. A sound's wavelength translates directly into how far along the cochlear spiral the wave breaks, exciting the hairs. High-frequency sounds release their energy by moving hairs early on; lower-frequency waves stimulate hairs farther along.

Finally, the hairs convert their movement into electrical signals and send them on to the brain. And it all happens in an instant.

Echo, Echo When we explored light and its wavelengths, we were dealing with extremely small fluctuations in electromagnetic radiation—sizes on the order of millionths and billionths of a meter. Sound waves are far longer. Even the highest pitch we can hear—the one with the most vibrations per second and therefore the smallest waves—represents a wave about 1.7 cm long, from crest

> "There's one thing I hate! All the noise, noise, noise, noise!"
> —Dr. Seuss, *How the Grinch Stole Christmas*

> "The world is never quiet, even its silence eternally resounds with the same notes, in vibrations which escape our ears."
> —Albert Camus, *The Rebel*

to crest. That's about 20,000 times longer than the longest visible electromagnetic wave we can see, red light!

You can easily calculate a wavelength by dividing the speed of sound by the frequency of the sound. So the musical note middle C, with a vibrational frequency of about 262 Hz, corresponds to a wave about 4.3 feet (1.3 m) long. The lowest pitch humans can hear is an astounding 75 feet (22 m) long.

Of course, traveling at 343 m/sec, these long waves, with their alternating increase and decrease of air pressure, still roll over us in a matter of milliseconds.

The length of a sound wave affects how we hear it in another way, too. When a wave of any kind hits an obstacle, the result depends on the wavelength. A wave shorter than the obstacle tends to reflect off it. We can see that easily with light waves, far smaller than even the tiniest object we can see with our eyes. Shine a flashlight on something and it causes a shadow where the light does not reach.

The same thing happens with sound waves: You can shout close to a wall and hear its reflection. Submariners take advantage of this effect, navigating the murky depths with sonar—flashes of sound that echo back the location of large objects in the water.

But due to their size, sound waves don't reflect off small things— anything smaller than the size of the wave itself. Instead, they diffract—they bend around. That's why you can hear music from a stereo down the hall, though it tends to sound muted with too much bass. The higher-frequency short-wavelength sounds get blocked at doorways and don't diffract as well. Lower-frequency sounds—those with longer wavelengths—tend to bend around corners quite well, allowing them to travel far and wide.

This is also why you can close your eyes and turn your head left to right and you are able to locate where in the room someone is speaking: Your own head literally makes a sound shadow, and the higher frequencies of the voice go in one ear more than the other. Similarly, stereo enthusiasts know that it doesn't matter where in a room you put a low-frequency subwoofer; the huge wavelengths

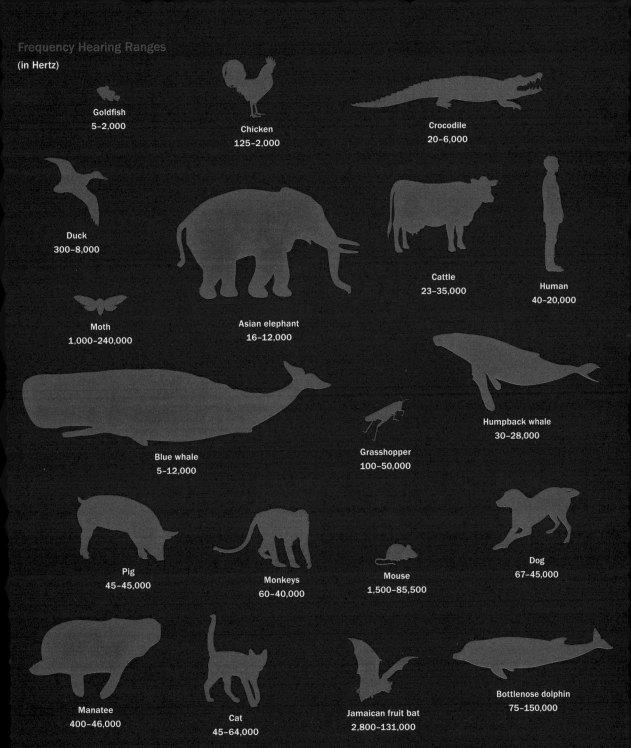

Frequency Hearing Ranges
(in Hertz)

Goldfish
5–2,000

Chicken
125–2,000

Crocodile
20–6,000

Duck
300–8,000

Cattle
23–35,000

Human
40–20,000

Moth
1,000–240,000

Asian elephant
16–12,000

Humpback whale
30–28,000

Blue whale
5–12,000

Grasshopper
100–50,000

Pig
45–45,000

Monkeys
60–40,000

Mouse
1,500–85,500

Dog
67–45,000

Manatee
400–46,000

Cat
45–64,000

Jamaican fruit bat
2,800–131,000

Bottlenose dolphin
75–150,000

diffract so much that you're unlikely to be able to tell if the source is in front of or behind you.

There's another reason sound quality changes through space: Matter (the air, your chair, whatever) absorbs sound energy, converting it into a hardly detectable amount of heat. A large room or a deep canyon may echo every sound, but the quality of the tone tends to be somewhat muted because higher-frequency smaller-wavelength sounds are absorbed more quickly than those low-frequency thumpers. This explains why a thunderclap sounds like a sharp crack when it's near you but only a low rumble a mile or two away.

Beyond Our Sound The fact that as we age we naturally lose the ability to hear higher frequencies may make you wonder if there are other sounds "out there" that you're not hearing. The answer is absolutely, though they're not necessarily sounds you want to hear.

Sound waves at frequencies higher than 20 kHz are called ultrasonic (that's different from supersonic, which means faster than sound). Dogs can hear ultrasonic vibrations up to 45,000 cycles per second, and cats probably hear a bit beyond that. The reason is likely evolutionary: If you're hunting a small animal like a mouse, you want to hear it—and the call of a young mouse in distress can easily hit 40 kHz.

Hunting by sound is also the key to a practice known as echolocation. If you can make a sound wave narrow enough to reflect off whatever it is that you want to eat, you can find it—sense it, almost feel it—by listening to the echoes around you. The classic example is bats, who can bark at well over 100 kHz. The sound is very loud and very short (usually only a few milliseconds in duration), but the wavelength is just right, bouncing off anything as small as 2 or 3 millimeters—quite helpful when looking for insects or avoiding the branches of trees. When it comes to larger animals or objects, bats can even "see" small features, letting them know what something looks like, what kind of animal they're approaching, and so on. Even more astonishing is the fact that bats often fly in packs of hundreds or even thousands but can still navigate by recognizing their own voices.

Humans have found a number of clever ways to make use of ultrasound. Dentists use it to clean teeth, doctors focus it to break up kidney stones noninvasively (a practice known by its tongue-twister name *lithotripsy*), therapists use it to apply "deep heat" muscle treatments, and diagnosticians and engineers use ultrasonography to visualize the interior of the human body as well as test the structural integrity of plastics, wood, or metal beams. However, these practices all use sound waves with frequencies far beyond those audible to animals, with waves reaching from 50,000 cycles per second up to 18 million hertz (18 MHz).

At the other end of the sound frequency spectrum, below 20 waves per second, sits the mysterious world of infrasound. Whereas dolphins, porpoises, and orca whales echolocate in the ultrasonic range, when it comes to communication, most marine mammals tend toward this lower end of the sound spectrum. As we've seen, lower-frequency sounds can travel farther, and those below 1,000 Hz happen to travel much farther in the saltwater sea. So the humpback and blue whale can sing out extremely loudly (over 150 dB) in the range of 10 to 30 Hz—low, powerful songs that can travel hundreds of miles.

On land, elephants, hippopotamuses, and alligators also use these infrasonic tones (sounds with a frequency too low for humans to hear) to communicate with their brethren, allowing for widespread coordination of herds, or for males to find mates. A female elephant, for example, signals its availability by creating distinctive rumbling noises that can broadcast over several kilometers. Zoologists report they can feel these calls thrumming through the air even when they cannot hear them. It's unclear how the animals themselves detect

Humpback whales sing songs so loudly that they can be heard more than 100 mi (160 km) away. However, it is sperm whales that emit the loudest sound of any animal, using an air-blowing structure in their heads, curiously named "monkey lips." These clicks can be as loud as 230 dB!

these noises, though as pressure waves travel better through rock than air, it's possible that they feel the sound through their feet.

Curiously, even though we humans cannot hear infrasonic tones, we can detect them, and the effects can be dramatic. In 2003, a team of researchers in London set up an "infrasonic cannon" in the back of a concert hall, adding very soft (only about 7 dB) and very low (17 Hz) sounds intermittently to several pieces of music played before a large audience. When asked, 22 percent of the listeners reported feelings of intense discomfort, fear, or—on the flip side—a sense of the supernatural or numinous, using phrases such as "an odd feeling in my stomach," "feeling very anxious," and a "strange blend of tranquility and unease."

The fact that infrasound makes humans uncomfortable has been repeatedly documented. Employees have refused to work in certain factory rooms in which they felt inexplicably ill, until it was discovered that vibrating cooling fans were pumping more than just air into the environment. Some scientists now believe that many reports of haunted houses actually stem from underlying and hard-to-trace infrasonic waves. For example, a sound wave at just the right frequency—about 18 Hz—can actually cause the human eye to vibrate, and this vibration may cause mysterious gray apparitions in the peripheral vision. Could ghosts be what Shakespeare's Macbeth called "full of sound and fury, signifying nothing"?

The lowest sounds ever found—far below the infrasonic thunder of animals, avalanches, or earthquakes, measured at even less than 1 hertz—are those of distant cosmic events. Many galaxies contain a huge quantity of "free-floating" gas—the residue of untold numbers of stars that have grown and exploded over billions of years. Astronomers, looking at a black hole in the Perseus cluster of galaxies, about 250 million light-years away, recently noticed a pattern in the clouds. More dense in some areas, less dense in others, the pattern soon revealed itself to be a sound wave, emanating from the black hole.

Many species have developed specialized organs to produce or detect sounds. Arthropods such as spiders and cockroaches have special hairs on their legs that can sense sound. The antennae of mosquitoes and many other insects can sense minute variations in air pressure. Honeybees appear to communicate through the buzzing vibrations of their wings; ants, crickets, and even some snakes and spiders stridulate (rub body parts together) to create chirps, clicks, or hisses. Some of these can be extremely loud: The African cicada's sound has been measured at over 105 dB!

The sound is a single note, drawn out not over meters but over billions of meters. To be precise, the researchers have determined that it's a B-flat 57 octaves below middle C, a million billion times lower than the lowest sound we can hear. If you can imagine, where a 20 Hz sound wave would take 1/20 of a second to pass by, a single wave of the Perseus black hole's drone would take 10 million years. Truly, the irony is palpable; Plato noted that "the empty vessel makes the loudest sound."

Of course, as we've seen, sound is ultimately absorbed and converted into heat. And scientists estimate that these tones, these rich, vast roars, provide as much energy throughout a galaxy as billions of suns. It's as though the music of the spheres, heating this interstellar gas, helps create just the right conditions for new stars and galaxies to be born.

Complex Sounds Today, on our small planet, we are saturated with an astonishingly rich sound field—one in which the spectrums of frequency and amplitude are woven together alongside rhythm, cadence, harmony, and many other audible ingredients. Even more amazing is that we can make sense of the melee. You may be in conversation with someone at a dinner party, overhearing another discussion at the table, tapping your toe to the music in the background, and suddenly catch the cry of a baby in the next room.

One reason we can distinguish among these various signals is that sound is rarely pure. If a flute and a piano each played a slightly flat A by emitting a perfect 436 Hz frequency, we'd never tell the difference between them. But musical instruments, and most objects that produce a sound, create overtones—combinations of additional frequencies, usually at even multiples above the fundamental, lowest note, called harmonics. So play a flute—essentially a metal tube with some holes in it—at 436 hertz and you'll invariably find a strong second tone at 872 hertz added to the mix, along with a dash of 1,308 Hz and 1,744 Hz. You'll also hear a faint jumble of many other frequencies between, creating the characteristic breathy sound. A

In the beginning there was silence, and then, suddenly, there was light. When the astronomer and science-fiction writer Fred Hoyle coined the term "big bang" in 1949, he didn't intend to describe the sound of creation, and in fact it's a misnomer—with no medium, the explosive expansion of the new universe would have been a light show without a sound track. But it wasn't long before vibration began to pulse through the blinding sea of photons, and then later the primordial soup of early atoms. The effects of these earliest waves can be detected even today, 13.7 billion years later, as astronomers map the heavens. We can see galaxies—each filled with billions of stars—clumping together every 500 million light-years or so in alternating crests and troughs, compression and rarefaction.

piano combines the same ingredients in very different amounts to concoct a completely different flavo The exact blend of frequencies helps define what we call the instrument's tone, or timbre.

Nevertheless, the processing required to decipher the billions of overlapping waves we hear during a few moments at that dinner party seems beyond possibility. Yet it's just another day at the office for our ears.

And, it should be noted, for our skin—for even the deaf can sense and appreciate sound waves. Many deaf people enjoy dancing to the rhythms of loud, rumbling bass-l that they can feel. At music concerts, some deaf audi hbers hold balloons between their fingertips that act like external eardrums, resonating and amplifying the sound, allowing a richer appreciation.

While at first glance this may seem like a completely different activity from hearing, remember that hearing *is* feeling. In fact, brain researchers have recently discovered that deaf people experience these physical vibrations in the same part of the brain where hearing people process sound from the ears. It's clear that humans are designed to detect and understand sound, one way or another.

There is no doubt that sound is a crucial part of our human experience, and perhaps even beyond. Virtually every faith tradition focuses on the creative and healing power of sound. In the Sufi Muslim tradition, music and chant are the secret of bringing one closer to Sirr, the center of inner consciousness where contact with the Divine is possible. As the influential twelfth-century Islamic philosopher Abu Hamid al-Ghazzali said, "There is no way to the extracting of [the heart's] hidden things save by the flint and steel of listening to music and singing, and there is no entrance to the heart save by the antechamber of the ears."

Similarly, Christians and Jews focus on the "Word" of spiritual revelation, reflected through the prophets or—in the mystical traditions—our own chants, purportedly generating heavenly reverberations. Eastern religious practices include meditative intonations of mantras, often based on the "universal sound" of *aum* or *om*. This is an ancient idea—that sound can be both viscerally and metaphysically transformative, resonating within ourselves and to the celestial heights.

And the idea is certainly not without merit, as the phenomenon of resonance is easily shown. Every object, every material, has a natural frequency at which it vibrates. For example, tink the side of a wineglass to hear its special pitch. If you play a matching tone, your sound's waves add to the object's—make it loud enough and you can cause the material to shake to the point of breaking. The stories of singers shattering glass good are true!

You can see this trick of pi̇ any kind of wave. Pushing a child in a swing at intervals that match the swing's resonant frequency makes it go higher, even with very little effort. Push at a faster or slower rate, and neither you nor your child will enjoy it. A piano or cello string does the same thing all by itself: Play a similar note on another instrument—either the same pitch or one that shares the same harmonic overtone—and the string sings out in reply.

So who is to say if the sound of song or prayer could not excite that which is beyond our seeing?

As our ears capture signals from the waves, we become aware of our interconnectedness with the matter around us. A footstep, a word, our favorite song, a brush of silk—we discover significance in the sounds that travel to us and through us, just as we create sound to convey meaning and relationship. These cycles of energy are powerful influencers, from the dark rushings of the mother's womb to the explosive death of stars.

HEAT

It doesn't make a difference what temperature a room is, it's always room temperature.

—Steven Wright

THERE ARE FEW THINGS MORE PLEASURABLE, ON EARTH OR IN the heavens, as a good hot bath or shower. Its enveloping warmth is reassuring, replenishing, rejuvenating—and for a good reason: Heat implicitly means life. Life requires heat, though like the fable of Goldilocks's bears, not too much heat, not too little, but just the right amount. The heat from a bath or a lover's embrace is a reminder that life is good and, if only for a moment, all is well.

Heat, at its essence, is motion—the motion of atoms and molecules goaded into movement by electricity, compression, chemical reactions, nuclear forces, or one of many other sources of energy. Energy inevitably turns into motion, like kids on a sugar high, and the motion spreads from atom to atom, molecule to molecule, until the warmth is shared as equally as possible.

Temperature—the measure of heat—is our natural way of gauging how much energy is in something. An ice cube has little energy to offer. A sweltering hot day is buzzing with energy, though the ambient humidity may make you feel as though you're drained of yours. We are constantly aware of temperature, because every aspect of our life and health depends on it.

And yet we actually recognize only a small sliver of the wide range of temperatures in nature. Our fragile bodies can handle only the smallest variation in heat. An object only 30 degrees greater than our own internal temperature can cause significant burns, and if our own body temperature drops even 10 degrees for any significant amount of time, the result is catastrophic. Of course, our warm-blooded metabolism lets us regulate our body heat appropriately, so we perspire to cool ourselves, or generate cell heat as required, even resorting to the wild movements of involuntary shivering if necessary. But if these systems fail, either hyper- or hypothermia can set in, shutting down key chemical reactions in your organs and ultimately leading to death.

Other animals have adapted to deal with heat fluctuations in other ways. The North American wood frog doesn't even try to get warm when winter sets in. Instead, as the temperature drops, it suffuses its cells and bloodstream with a cocktail of sugars and proteins that allows it to freeze solid without tissue damage. Once frozen, it shows no signs of life whatsoever: no heartbeat, no breathing, no kidney function. It is as dead as a stone . . . until the spring thaw, when some deep unknown signal miraculously tells everything to start up again, and in a matter of hours the frog is hopping about looking for a mate.

We've learned to survive in the harshest of both arctic and desert conditions, but even these are temperate compared with some places in space, or deep inside the crust of our planet. Temperature is energy in motion, and energy—as you may guess—covers a wide gamut.

> **A bolt of lightning can reach** 50,000°F—hotter than the surface of the sun—and packs a punch between 100 million and 1 billion volts.

Measuring Temperature If you were traveling to Sweden and heard the temperature was 22°, would you want to bring a jacket? Would you worry if, instead, the forecast read 295°? It all depends on the scale you're using, of course: Celsius, Kelvin, or Fahrenheit.

Scientists as early as the second century BCE discovered that certain substances, such as air, expand when heated, but it was not until the seventeenth century that scientists such as Galileo Galilei used this knowledge to build devices that would measure heat itself, called "thermo-meters." When Isaac Newton made the radical suggestion to place marks on the thermometer in order to better record specific values, he prescribed that the zero mark

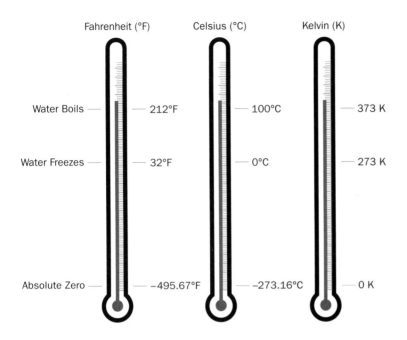

Fahrenheit (°F)   Celsius (°C)   Kelvin (K)

Water Boils —   — 212°F   — 100°C   — 373 K

Water Freezes —   — 32°F   — 0°C   — 273 K

Absolute Zero —   — −495.67°F   — −273.16°C   — 0 K

should indicate melting ice and 12 reflect the temperature of the human body.

The value of 12 may seem odd (why not 10 or some other reasonable number?), but note that this duodecimal system is particularly handy when it comes to splitting evenly into sixths, quarters, thirds, and halves. Thus, the English standardized on 12 inches to a foot, 12 pence to the shilling, 12 units in a dozen, 12 dozens in a gross, and so on.

In 1714, a young glassblower-cum-physicist named Daniel Gabriel Fahrenheit hit upon several genius improvements. Instead of using sticky, imprecise liquids such as alcohol inside the thermometer, he used mercury. In order to encompass a wider range of values, he set the zero mark at the melting point of saltwater, which freezes at a significantly lower temperature than freshwater. And he enabled finer increments by increasing the top value, the body's temperature,

to 96°. (Once again, the choice of 96 makes sense only when you notice that it's easily divisible by 2, 3, 4, 6, 8, 12, and so on.)

A few years later, as scientists decided the boiling point of water was a more important value than body temperature, Fahrenheit's scale was fudged slightly. Water, it was declared, should boil at exactly 180 degrees above its freezing point of 32°, at 212°F. This adjustment was convenient (180 is also easily split into smaller fractions and matches up nicely with the number of degrees in a half circle), but this required stretching Fahrenheit's scale a little bit, so that our body temperature would now be marked at the awkward value of 98.6°F.

Given that the melting and boiling points of water are such important measurements (at least here on Earth), why not set those values to 0 and 100? Such was the reasoning of the Swedish astronomer Anders Celsius in 1742. (Actually, to be accurate, he bizarrely set freezing at 100 and boiling at 0, but that was quickly rectified a couple of years later, soon after he died.) Because each division on this thermometer measured exactly one hundredth of the total scale, it was labeled as a "degree centigrade" (Latin for "hundred steps"). That nomenclature stuck for three hundred years, until, in 1948, the term was changed by the International Committee for Weights and Measures to "degrees Celsius."

Now that we had two different ways of measuring heat, why bother with a third? By the mid-eighteenth century, scientists had realized that there was a world far beyond that of the boiling and freezing of water. At first, there didn't appear to be any limit to how hot or cold a substance could get. After all, if you could heat something to 1,000° Celsius (the temperature in a typical stovetop flame), then why not cool it to −1,000°C?

Unfortunately for this theory, as people were finding clever ways to cool nitrogen and other gases to their freezing point, they discovered a curiosity: For each degree Celsius you cool a gas, it reduces in volume a tiny amount—about 1/273. That led them to an

> **To convert Fahrenheit to** Celsius, subtract 32, multiply by 5, and then divide by 9. To convert °C to °F, multiply by 9, divide by 5, and then add 32.

intriguing conclusion: At that rate, the gas would disappear entirely, or at least take up zero space, when it reached –273°C. While their understanding of the science was still immature, they did correctly surmise that this must signify the coldest temperature possible.

The data was compelling enough that the Scottish physicist William Thomson—who had gained fame and a knighthood for his work on the telegraph—suggested that this lowest low should be the new zero, an *absolute* zero. Sir William's idea stuck, but you never hear about "degrees Thomson." Rather, he was later raised to the House of Lords, assuming the heady title Lord Kelvin of Largs. (He lived in Largs, and his office was on the river Kelvin, which flows through Glasgow.) So scientists began to talk in terms of "degrees Kelvin"—a system in which each degree is the same "size" as a degree Celsius, but in which zero starts much lower.

Because heat is motion, and motion is energy, and there are many different forms of energy and reasons to discuss them, science has developed a number of other ways to describe and discuss heat in a system. We talk about calories, for example, and when discussing the heat from a burning gas we might talk about joules or numbers of BTU (British thermal units), where a single BTU describes the energy given off by a wooden kitchen match.

Fortunately, we don't need this alphabet soup of heat descriptors to talk about how hot a proper cup of tea should be, or the weather on Venus. For common, daily usage, degrees Fahrenheit or Celsius do fine, and even when exploring the very cold or very hot, all but the geekiest discussions can use kelvins.

The Chaos Meter Here's an amazing magic trick you can try at home: Place a chunk of ice on a plate and leave it out on the counter. By mumbling numerous incantations (and just waiting awhile) that solid object transforms into—gasp!—a liquid. Wait long enough and that clear puddle—double gasp!—appears to vanish completely. Like any good magic trick, it seems miraculous until you know how it's done,

▲ Sir William Thomson, later Lord Kelvin

Scientists attending the 1967 General Conference on Weights and Measures gave Lord Kelvin one of the greatest honors of all when they agreed to drop the word *degree* and just call the measurements "kelvins." Thus, he joined a select group of inventors whose names have become uncapitalized measurements, such as watt, volt, ampere, and joule.

at which point it becomes ordinary. But allow yourself a moment to see the magic through fresh eyes. A solid turning liquid turning gas, a seemingly insignificant element: heat.

For centuries, scientists assumed that heat was literally an element—an invisible fluid called "caloric" that traveled from object to object. The seventeenth-century English philosopher and father of liberalism John Locke suggested that heat was a form of kinetic energy—the motion of the tiny "insensible parts" of a substance.

If heat is motion (which we now know it is), then there is technically no such thing as "cold." Obviously, one object may feel colder than another, but what you're really talking about is heat—there is only heat, which can be added and removed, motion sped up or slowed down. In other words, one object isn't really colder than the other; it's just less warm.

And here's the clever part: The measure of heat is also the measure of how much chaos or order there is. Cooling a substance removes energy, allowing molecules to rest into rigid structures. Adding heat smashes up that architecture, leaving a somewhat dense broth; adding even more heat allows the molecules to break free from one another, a rapidly expanding gas of entropy like a flock of birds rushing into the sky. Even the word *gas* itself stems from a Dutch pronunciation of the Greek word *chaos.*

It's important to remember that at the atomic level, nothing ever stops moving. On a pleasant spring day, air molecules are flying at 1,150 mph (1,850 km/h), manically bouncing off one another's magnetic fields, gently buzzing your skin, generating pressure and transporting heat. Even in a solid, like a crystal in which each atom is held tightly in place, molecules never stop dancing, wriggling atom to atom along their internal degrees of freedom. And, in turn, each atom vibrates with activity, each electron a never-ending now-it's-there-now-it's-not blur of enterprise.

One result of this constant motion is that solids aren't always as solid as we think they are, and molecules that you thought were pretty consistent sometimes unexpectedly change from one phase to another. For example, some of the molecules in a block of ice will, even if kept below freezing, change into a liquid phase, then usually freeze again. Even stranger, if left alone, the ice will eventually evaporate, as frozen molecules literally boil into vapor in a phase-skipping process called sublimation. Similarly, in a gas, a few cooler molecules will spontaneously combine to form a liquid, then return to gas again, in a continuous state of transmutation.

The most you can say about any material is that it tends toward a particular phase (solid, liquid, or gas) at a particular temperature and a particular pressure—for pressure, too, has a huge effect on phase. Increasing pressure increases temperature—that's why a bicycle tire gets warmer as you pump it full of air—but it also changes the melting or boiling point of a substance. Water boils at a lower temperature on top of a mountain than in a valley, as there is less air pressure to hold it in liquid form. Take a liquid far higher, into the low-pressure vacuum of space, and it instantaneously vaporizes, expands, cools, and then sublimates into tiny shards of crystal.

One thing is for certain, though: However matter changes, heat is never actually gained or lost; it is simply moved from place to place, or transformed from one kind of energy to another. This is at the heart of the laws of thermodynamics (a fancy way of saying "how heat moves"). Just as a gas always expands to fill its container, or people spread out to fill the space in an elevator, heat always spreads into cooler areas. Thus, when you use a thermometer to take your temperature, your body heat cools down a tiny bit as energy moves into the probe, until the two (body and device) are the same.

Each degree on the Fahrenheit scale equals ⅝ of a degree on the Celsius scale. That has the curious result that both scales converge at −40 degrees: −40°C = −40°F.

Making Things Cold Few people understand how a refrigerator—or an air conditioner, for that matter—makes the air cool. You can't just take air and "add cold" to it; you need to suck the heat out of the air you already have. When you use a spray can, you notice something curious: The longer you spray, the colder the can gets. The reason is simple: The pressure inside the can drops when you press the button, and the lower the pressure, the less the molecules bounce around, and so the colder the gas becomes. Of course, after a moment, the can absorbs heat from the air around it, so the effect fades.

A refrigerator captures this same effect but carefully controls it— and keeps the chemicals in a closed loop in order to use them again. A substance such as liquid carbon dioxide or Freon is released from a compressed tube, through a spray nozzle, into a larger set of tubes, where the gas rapidly expands and becomes very cold, very quickly. Air from inside the refrigerator is blown over these tubes, transferring any heat to the colder-than-ice gas. The air circulating around your food (or room) gets colder, and the now warmer gas is then pumped out, compressed until it becomes very hot, and run through condenser coils, where all the heat is released into the room (or outside). By the time the gas gets to the end of these coils, it has returned to room temperature (and much of it has turned to liquid),

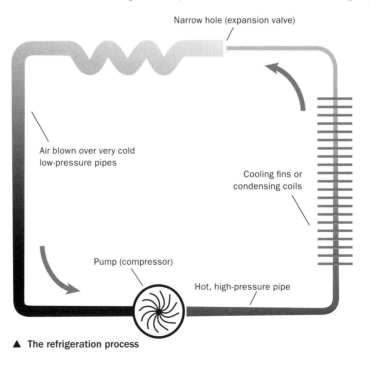

▲ The refrigeration process

but it's still highly compressed, ready to repeat the process all over again.

Meanwhile, while you're cooling your food or home or whatever, you're also pulling water out of the air—dehumidifying it. The cooler the air, the less moisture it's able to contain, so water vapor collects (called condensation) around the cold pipes—just another phase change due to heat.

The funny thing about water (which makes up the vast majority of us, the food we eat, and the surface of our planet) is that it expands slightly when cooled. This is unusual, as most other substances get more dense as they cool and freeze. This property of water is one reason life does so well on Earth—after all, the bigger, less dense ice floats instead of sinking, letting water freeze from the top down, so fish and other plants can live protected from the cold.

Unfortunately, even though ice takes up only 9 percent more space than water, that's enough to cause massive destruction where you least want it. When water leaks into a tiny crack and freezes, the force of the expansion can split rock or concrete; metal pipes can burst, glass can shatter. Organic materials like meat and vegetables—which you'd expect to be more pliable—often fare the worst, as sharp ice crystals rip open delicate cell membranes, rupturing their contents. The result, when thawed, is the flavorless, mushy mess that disheartens so many hopeful cooks.

In 1923, the inventor Clarence Birdseye found that quickly freezing a thin layer of fish filets created much smaller and more evenly spaced ice crystals that avoided most of the cell damage. What works for fish works for other food, too, and by 1928, Americans were buying over a million pounds of frozen foods each year. In theory, flash-frozen food will hold indefinitely, but as we've learned, even ice cubes evaporate, and moisture slowly wicks away wherever cold air can get to it, causing unsightly (and tissue-damaging) freezer burn.

Of course, if you want to freeze something really fast, you put it in an environment far colder than normal ice. Take frozen carbon

Because cold air contains less moisture, mountain ranges in temperate areas receive far more snow than the north or south poles. The Antarctic is essentially a desert, with extremely dry air and even less precipitation than Phoenix, Arizona!

Denizens of cold climes know that stepping on snow causes little noise at air temperatures near 0°C, when a thin film of water lubricates the rubbing between the ice crystals. At much lower temperatures there is no water film, so the friction produces a relaxation oscillation called a squeak.

dioxide (CO_2), for example, otherwise known as dry ice. The French scientist Charles Thilorier was the first to change the phase of this gas, in 1834, by placing it under incredible pressure (thus radically raising its boiling point), then releasing the pressure, dropping the temperature so quickly that it desublimated directly into a solid. It sounds like a simple process, but at the time, these kinds of experiments were dangerous; one of Thilorier's assistants lost both legs when the apparatus exploded during testing.

You can buy dry ice at a grocery store, but don't touch it: The surface is −79°C (−110°F), so cold that it burns, destroying your skin

Element	Melting Point (°C) (solid to liquid)	Boiling Point (°C) (liquid to gas)
Helium	−272	−269
Hydrogen	−259	−253
Oxygen	−223	−183
Nitrogen	−210	−196
Chlorine	−101	−35
Carbon dioxide	−78	−57
Mercury	−39	−357
Bromine	−7	59
Phosphorus	44	280
Lead	328	1,740
Aluminum	660	2,467
Silver	961	2,212
Gold	1,065	2,807
Iron	1,535	2,750
Tungsten	3,422	5,555
Carbon	3,550	4,827

▲ Phase change

cells. However, it'll keep a container cold for quite a while as the solid CO_2 slowly, dryly evaporates back into a somewhat harmless gas.

The Hunt for Zero Frozen carbon dioxide's melting point of –79°C is pretty nippy, but in 1983, at the Vostok Station in Antarctica, the thermometer outside dropped to –89.2°C (–128.6°F)—currently the coldest naturally occurring temperature ever recorded on Earth. There have certainly been events even colder, but without the benefit of humans to record them.

At –100°C, rubber tires freeze—a point well taken by recyclers, who shatter them into tiny shards to be reused in other materials. About 80 degrees colder, the oxygen and nitrogen we breathe liquify; and only 40 degrees colder than that, at –219°C (–362°F), they turn solid.

Liquid nitrogen and oxygen have a wide array of uses, as they freeze almost anything virtually instantaneously, perfectly preserving it in stasis. Cryobiologists commonly keep sperm cells, stem cells, and many other plant and animal tissues at –196°C (–320°F) indefinitely with little loss when thawed. And if you can freeze a cell, why not an organ, or even a whole body? In the early 1960s, a Japanese researcher froze a number of cat brains for days, weeks, and even months. After they were carefully warmed with a bloodlike substance, there were measurable (though brief) brain signals that were very similar to those from the original live brain.

Following this promising (though disquieting) evidence, cryonics experts have frozen hundreds of human bodies in liquid nitrogen in hopes that someday technology will advance enough to reanimate these people. Some customers choose to preserve only their brains (removed as quickly as possible after the moment of death), assuming that any future civilization capable of thawing a whole body could just as likely transplant the brain into a new body, or perhaps even create a new brain using the original, intricately woven neurons as a model.

Another benefit of compressed liquid air is that, when released, it boils from a compact form to a gas extremely rapidly—even explosively. In 1926, the American physicist Robert Goddard found a way to control this reaction, turning it into a fuel to propel a rocket—the same technique NASA and every international space agency has used to launch satellites, place astronauts on the moon, and boost shuttles to the International Space Station. Clearly there is great power in carefully managing the cold.

Space can get far colder than our temperate little planet, of course. The surface of Pluto is about −223°C (−369°F), and a crater on the dark side of the moon has been measured at a few degrees colder than that. While it would seem like there would be no temperature at all in the dead of space, far out between the galaxies where no stars burn, astronomers have discovered that even "emptiness" has an amazingly consistent "background radiation" temperature of about −270°C, or −455°F—more easily notated as 2.7 kelvins. This ubiquitous but mysterious heat has fascinating implications for our understanding of how our universe was born and grew.

Shortly after the big bang (approximately 13.7 billion years ago), the universe contained a lot of material in a still relatively small space and glowed unbelievably brightly with extreme heat. Astrophysicists believe that the faint background radiation, which they can detect no matter where they point radio telescopes, is the remnant of those early days, like the heat on an oven's walls long after the oven has been turned off. Curiously, on closer inspection, they've found that space is a tiny bit warmer in some places and a tiny bit cooler in others—for example, it may be 2.7249 K in one spot and 2.7250 K in another. These infinitesimal differences are likely the result of quantum fluctuations that have been stretched out by the inflation of the space-time continuum. In other words, as the universe has cooled and expanded, the massive differences between scorching and less-scorched spots that must have existed within the explosive fireball have all cooled and almost evened out, creating this pattern.

These temperatures are cold enough to freeze almost anything except helium, with a melting point of 0.95 K (–272.2°C, –458°F)—which scientists long considered the final frontier. Then, in 1908, the Dutch physicist Heike Kamerlingh Onnes accomplished the task in an experiment that took seven years of preparation, followed by thirteen hours of slowly cooling the helium until it solidified.

▲ The coldest known natural temperature in the universe is the Boomerang Nebula, a bow-tie-shaped expulsion of gas from a dying star about 5,000 light-years away in the constellation Centaurus. You'd expect an explosion like this to generate huge quantities of heat, but in fact the extremely rapid expansion of the gas into space had just the opposite effect, acting like the expansion tube in a refrigerator and leading to a temperature of about 1 K.

Referring to the explorers who were attempting to reach Earth's poles during that same decade, Kamerlingh Onnes explained his passion for this work: "The arctic regions in physics incite the experimenter as the extreme north and south incite the discoverer."

Where It All Gets Wacky Reaching toward absolute zero is like one of those horrible nightmares where the faster you try to run, the slower you get. We know that zero kelvin (−273.16°C) is a hard stop we can never achieve—a physical extremity like the speed of light. At absolute zero, all atomic and subatomic motion would stop and all particles would come to rest with zero energy. As far as our current understanding of science can see, accomplishing absolute zero would violate one of the fundamental laws of quantum physics, Heisenberg's uncertainty principle, which states that we cannot pinpoint the exact location and momentum of every particle, even electrons.

This kind of boundary—where each step closer is exponentially more difficult—is called an asymptotic limit. In this case, as you get closer to zero, that is, as you are trying to remove more and more heat, you actually start generating heat. But the challenge is worth the effort, for the world of the ultracold reveals some astonishing phenomena.

To play in this realm, you need more clever ways of reducing heat besides just compression and expansion. For example, common evaporation lowers surface temperature, which we humans take advantage of by sweating on a hot day. You can also suck heat out of a system by running a small electrical charge to create a difference in temperature between two metal plates, a device used in many portable camping coolers. But when working at extremely low temperatures, below 1 K, nothing compares to cooling atoms with lasers.

Laser cooling techniques, developed in the mid 1980s with names like Doppler or Sisyphus cooling, all work by focusing two or more intense beams of light at a tiny group of atoms. By precisely tuning

Temperatures

Absolute hot (Planck temperature) 1.42×10^{32} K

Melting point of hadrons into quark-gluon plasma 2 trillion K

Everything 1 second after big bang 10 billion K

Thermonuclear weapon peak 350 million °C

Sun's core 15 million °C (27 million °F)

Lightning bolt 28,000°C (50,000°F)

Center of Earth 6,650°C (12,000°F)

Surface of sun 5,500°C (10,000°F)

Filament inside light bulb 2,500°C (4,600°F)

Natural gas (methane) flame on a stovetop 1,200°C (2,200°F)

Lava 1,100°C (2,000°F)

Wood fire 900°C (1,650°F)

Draper point (where almost all solid materials begin to visibly glow) 525°C (977°F)

Melting point of lead 328°C (621°F)

Kitchen oven 288°C (550°F)

Book cellulose-based paper burns 233°C (451°F)

Water boils 100°C (212°F)

Hottest shade temperature recorded on Earth 58°C (136°F)

Human body temperature 37°C (98.6°F)

Room temperature 20°C (68°F)

Water freezes 0°C (32°F or 273 K)

Mercury in a thermometer freezes –39°C (–38°F)

Coldest temperature recorded on Earth –89°C (–129°F)

Alcohol freezes –114°C (–173°F)

Gasoline freezes –150°C (–238°F)

Boiling temperature of oxygen –183°C (–298°F)

Temperature on Neptune –220°C (–364°F)

Coldest spot on moon –228°C (–378°F or 45 K)

Cosmic microwave background 2.725 K

Coldest natural temperature known (Boomerang Nebula) 1 K

Coldest measured temperature 100 pK

Absolute zero 0 K (–273.16°C or –495.67°F)

Note: Many values here are approximate, as temperatures can vary.

the electromagnetic wavelengths of the lasers, scientists can draw atoms in one direction or another by bombarding them with photons. One of the inventors of this method, the Nobel Prize–winning American physicist Carl Wieman, described the process as "like running in a hail storm so that no matter what direction you run the hail is always hitting you in the face . . . So you stop."

In the 1980s, scientists achieved thousandths of a kelvin. In the '90s, they slowed down the atoms even more, lowering the temperature to millionths of a kelvin, then hundreds of billionths, offering an unprecedented glimpse at the world of the supersmall. Luis Orozco, a physics professor at the University of Maryland, explained in a *NOVA* documentary: "It is as if I were to ask you, 'Could you tell me something about the handles of a car that is passing on a highway at 50 or 60 miles an hour?' Definitely you won't be able to say anything. But if the car is moving rather slowly, then you would be able to tell me, 'Oh yes, the handle is this kind, that color . . .' At room temperature an atom is moving at roughly five hundred meters per second [about 1,100 mph]. However, if I slow it to a temperature that we can now achieve without much work in the lab, two hundred microKelvin, then the atoms start to move about twenty centimeters per second. Compared to something that's rushing in front of you, you'd be able to look at a lot of the details, a lot of the internal structure of that atom."

But it turns out that supercooling atoms encourages them to behave bizarrely—behavior that not only provides insights into the nature of matter but also may allow us to improve technology in extraordinary ways. For example, while some metals are better than others at conducting electricity, some materials, such as lead or buckminsterfullerines, become superconductors when cooled to extremes. A superconductor doesn't just allow electrical impulses to pass through it; it does so without offering any resistance—a current running through a loop of superconducting wire will never fade.

It's unclear how superconductors can pull off this feat of perpetual motion, but it appears that as the temperature drops, the atoms

If the sun went out, Earth's surface would cool to about –220°C, warmed slightly from heat coming from its core.

vibrate less and electrons can slip through more easily. The electrons actually seem to group into pairs, each tugging the other forward, when normally they would repel.

However it's accomplished, superconductivity has led to incredibly powerful and precise magnets—magnets that today power MRI scanners and particle accelerators. Magnetically levitating (maglev) trains based on superconductor technology are still being tested, but they have already broken the world record for fastest-moving train, on an experimental track in Japan, at 581 km/h (361 mph). Someday, whole power grids may be based on superconductors, as estimates suggest that, in transferring electricity, 110 kilograms (250 lb) of superconducting wire could replace 8,100 kg (18,000 lb) of copper wire.

A second characteristic of supercooled atoms is that they can become superfluid—that is, at a certain point, the atoms in liquid helium begin to ignore friction. If you swirl a superfluid, it keeps swirling forever; if you spin its container, the superfluid inside remains motionless. A superfluid can escape through extremely small pores that would normally hold any liquid. Weirdest of all, because a superfluid does have surface tension—like all liquids, it has a slight attraction to the sides of a glass container—it gradually creeps up the sides of a cup until it flows out on its own accord, like a translucent creature that somehow knows how to escape confinement.

But these superpowers were just a small taste of the wonders scientists were about to find in the nanokelvin zone. In 1995, when the physicists Carl Wieman and Eric Cornell cooled a small collection of rubidium atoms to these extremes, they encountered a breakthrough moment: The atoms suddenly shifted phase—not to a liquid, or a solid, but to an entirely new state of matter, never before seen.

To understand this new state, we need to look back to 1924, when Albert Einstein and the Indian physicist Satyendra Nath Bose theorized that as individual atoms neared absolute zero, they would change in an extraordinary way. Quantum mechanics states that all atoms can be described as either particles (things) or waves (energy). Near zero, the theory went, atoms should begin to act less like particles and more like waves; then the waves would get longer until they overlapped, suddenly acting like a single wave—that is, as though all the atoms were one "superatom."

This new state, a "holy grail of cold" called the Bose–Einstein condensate (BEC), is what Wieman and Cornell had created in their lab. Like everything else about quantum physics, BEC is a counterintuitive mystery. The atoms still exist, yet they have expanded their size—their awareness, as it were—in a way that we still don't understand.

The condensate behaves unlike any other material. The atoms vibrate—barely—in unison, a quantum lockstep that acts like a giant magnifying glass on what is usually far too small to see. In 1998, the Harvard physicist Lene Vestergaard Hau found that she could shine a laser into a BEC made of millions of sodium atoms and slow down the light to 68 km/h (38 mph)—a huge leap from its normal speed of 300,000 km (186,282 mi) per *second*. A few years later she found a way to tune the BEC using lasers of specific wavelengths, letting her literally stop the light entirely, then release it again on its way.

Here's how it works. The light pulse is converted to a hologram inside the condensate, literally creating a copy in matter, which can actually be transferred from one BEC to another nearby, like handing over a packet of information, before transforming back into light again. The implications are staggering and point to future quantum computers that may run on light instead of electricity.

On the other hand, these tiny condensates can also explode in an unexpected, extremely tiny version of a supernova—which scientists, recalling the 1960s Brazilian music boom, call a "Bosenova."

What happens in environments even colder than the nanokelvin? We're still finding out. After the Nobel laureate Wolfgang Ketterle trapped a cloud of sodium atoms in place with magnets in 2003, his team at MIT was able to laser-cool the gas to 500 picokelvins—half a billionth of a degree above absolute zero.

The physicist Juha Tuoriniemi at the Helsinki University of Technology's Low Temperature Laboratory has taken a small piece of rhodium metal as low as 100 pK (1×10^{-10} K), but every supercold researcher today is focused on the next breakthrough: the femtokelvin, millions of times colder than the temperatures required to build a Bose–Einstein condensate. Scientists are hungry to see what surprises await in this often unpredictable realm where our everyday assumptions are superseded by the improbable results of quantum physics.

Some Like It Hot It's a common misconception that heat rises. In reality, dense things sink and less dense things get pushed out of the way—which typically means they float up, like bubbles in a drink. That holds true whether it's air in a room or lava under Earth's crust. The difference in density is caused, of course, by heat. Add heat and most substances expand, lowering their density as the atoms and molecules dance and twitch. (We've already seen that ice offers one exception to this; silicon is another. But they are truly oddities among the vast majority of materials.)

Solids aren't in any position to move much, but everyone knows that running hot water over the metal lid of a tightly sealed jar makes it easier to open—the heat literally expands the metal, even if just a small amount. Engineers using steel in railroad tracks and bridges have to take this effect into account anywhere the ambient temperature is likely to rise or fall significantly. On a hot day, a beam may expand several millimeters in length, buckling if an expansion joint hasn't been provided.

A liquid or a gas offers plenty of latitude to move about, creating convection currents in which colder areas drop, get warmed, rise, lose some of their heat, and then drop again. This cycle is particularly helpful in cooking, but we can see it everywhere on Earth: weather patterns, the circulation of the ocean, hot soot rising up a chimney flue. Clearly, heat causes a lot of motion at the macro as well as the micro level.

At some point, if you add enough heat, you can force even more changes. A liquid boils, forcibly rending molecules apart until they fly into a gas. Even solids can change dramatically. If you place an intense heat source under a piece of paper, those pressed-flat plant fibers will undergo a radical transition: Around 150°C (300°F), the cellulose material starts to decompose, releasing into gases. We typically call this mixture of hydrogen, oxygen, and tiny carbon particles smoke. The more particulates that are released, the more smoky it appears. Some substances in paper don't burn without far more heat, of course, so some material remains, darkened, called "char."

If you apply even more heat, an astonishing thing happens: The various molecules in the paper and gas get so excited that they break apart into atoms, which quickly recombine to form carbon dioxide, water vapor, and other molecules. These blindingly fast chemical reactions have an interesting side effect: They generate even more heat, so even if you remove the original heat source, the new gases are so hot that they cause even more molecules to break up. As long as you have fuel to burn, and oxygen for it to react with, the heat keeps the process cycling. Obviously, we know this amazing chain reaction by a simple name: fire.

Fire has been held as magical for millennia, a gift to humankind so extraordinary that it must have been stolen from the gods, and assumed to be so fundamental that the ancients gave it elemental status alongside earth, air, and water. Now we know that fire is simply the transition from one state to another, and the flames we usually see are simply the gases and particles glowing with incandescent heat.

In fact, anything will glow when it gets hotter than 525°C (977°F), named the Draper point, after the nineteenth-century American chemist John William Draper, who first wrote about this effect. (Coincidentally, Draper was also fascinated by photography and is credited with producing the first clear photographs of a female face and of the moon.) To be accurate, anything cooler than 525°C glows,

The science-fiction author Ray Bradbury called his 1953 classic, about a man who burned books for a living, *Fahrenheit 451*. Of course, he could have named it the metric equivalent: *Celsius 233*. This is the temperature at which paper made from wood pulp begins to burn.

too, but with infrared light, so we can't see it. But at the Draper point the light waves carry enough energy that we begin to see a faint red glow. Heat something up to 725°C (1,337°F) and the color becomes positively luminous—red hot, as we call it.

It's easy to remember that red hot is about 1,000 K. At 3,000 K, a material glows bright orange; at 6,000 K, it turns yellow-white. Guess the surface temperature of our sun . . . right, just about 5,800 kelvins. If the sun's surface were hotter, it would appear bright white, or even—at 10,000 K—blue. This color spectrum is called black-body radiation, and it describes how thermal energy gets partially converted into electromagnetic energy, photons that will travel through space indefinitely. As long as there is movement, there is heat, and where there is heat, there is light.

Granted, some flames burn hot but we can't see them—or can barely see them. A pure hydrogen fire burns clear as the gas combines with oxygen in the air and turns into water vapor; a pure ethanol fire burns so hot that the blue is almost indistinguishable in the bright light of day. Plus, different chemicals release different colors when heated, explaining the rich tonal range of a wood fire. And sometimes the color of a flame doesn't tell the whole story: The blue section near the base of a match flame is technically hotter than the yellow tip, but due to a number of real-world factors (like air flow), it's generally easier to light a candle using the cooler tip.

Absolute Hot A candle flame is plenty warm enough for most of us, but it's only the beginning when it comes to the spectrum of hot. Because heat is just another form of energy, you can raise an object's temperature in all kinds of ways, from running an electrical current through it to blasting it with microwaves.

One way to get a gas hot is by compressing it, which also drops its boiling point so that it may return to a fluid state. If you apply enough pressure and heat, you create something called a supercritical fluid—not different enough to earn status as a new form of matter but nevertheless possessing some very cool properties.

"By convention sweet is sweet, by convention bitter is bitter, by convention hot is hot, by convention cold is cold, by convention color is color. But in reality there are atoms and the void. That is, the objects of sense are supposed to be real and it is customary to regard them as such, but in truth they are not. Only the atoms and the void are real."

—Democritus, Greek philosopher

For example, a supercritical fluid can dissolve materials like a liquid and also pass through semiporous solids like a gas. If you infuse a bunch of green coffee beans in a high-pressure bath of nontoxic hot carbon dioxide (CO_2), the supercritical fluid seeps through the beans, absorbing caffeine and drawing it out. Then release the pressure, and the CO_2 suddenly vaporizes into a steam, leaving the decaffeinated beans ready for roasting. Supercritical fluids are used in dry cleaning, essential oil extraction, dyeing materials, and all sorts of other applications.

However, if you warm up a gas even further, you do, in fact, achieve a new, fifth form of matter: plasma. At first, you may not notice any difference between a gas and a plasma, but the latter contains a bunch of atoms that have ionized—they've gotten so excited that they've broken free from their molecular relationships and stripped off some of their electrons, like partiers throwing caution to the wind.

This high-temperature concoction of positively charged ions and negatively charged electrons displays some interesting characteristics. To start with, you can run an electrical current through it, which is what makes neon signs, plasma televisions, and fluorescent lightbulbs glow. Switch on the power, and the molecules in the gas are turned into a superhot plasma. Fortunately, the gas is under such low pressure (there aren't that many atoms buzzing about) that the total heat enclosed in the lamp isn't hot enough to melt things around it.

In these applications, the color we see doesn't come from the heat that the lamps generate; instead, the plasma causes phosphorescent chemicals painted on the glass to luminesce. But in some other instances, the gaslike substance itself lights up as atoms and electrons reunite, resulting in bright light and intense heat. Plasma cutters, which spray a high-speed stream of electrified gas plasma through a nozzle, can cut through steel up to 6 inches (150 mm) thick.

We all know of another common plasma: the sun. In fact, all stars are made of plasma. And, weirder, most of the free-floating gas sparsely spread between planets and stars is in a plasma state. Astronomers estimate that as much as 99.9 percent of all the visible matter in the universe is plasma.

But just because something is superhot doesn't mean it's going to be plasma. Earth's core is hot—at 6,650°C (12,000°F), it's hotter down there than the surface of the sun—but gravity tugs on each atom across thousands of kilometers, creating intense pressure in the center, which is very hot but solid. The larger the mass, the more heat is generated, so Jupiter's core temperature is estimated to be as high as 20,000°C (36,000°F).

Yet that's a cold shower compared with what happens inside the blast of an atom bomb or nuclear reactor, where temperatures of millions of degrees can be achieved through fission—the splitting of heavy atoms like uranium into smaller ones. And fission, in turn, is somewhat pitiful compared with the universe's real power: fusion.

When a huge quantity of hydrogen and helium gas floating in space finds itself collected by gravity into a small enough ball, the atoms begin to smash into one another more and more rapidly. When the pressure and temperature are great enough, say, about 10 million K (about 18 million °F), the hydrogen atoms fuse together, nucleuses joining, creating a new helium atom. Sounds simple, but in the process a lot of energy is released—where "a lot" can be defined as mind-boggling, flabbergasting quantities that result in a fireball called a star.

The sun's core registers at about 15 million K (27 million °F), where hydrogen is consumed at about 600 million tons *per second.* That means the sun, which is only about 4.6 billion years old, is currently in its midlife, about halfway through burning out.

Fusion is not impossible on Earth, however. That's how the H-bomb works: A fission-based atom bomb blows inward on a small bit of prepared hydrogen, causing such intense heat that it fuses. The

Revelation 21:8 implies that Hell has lakes of brimstone (sulfur). At sea level, sulfur boils to gas at 444.6°C (832°F). However, far underground, at extremely high pressure, sulfur can stay liquid as high as 1,040°C (1,904°F).

▲ Large Hadron Collider

results, at temperatures reaching over 100 million K, are devastating. That said, if we can control fusion—or even better, create fusion at temperatures that don't require radioactive explosions to ignite—we would have a never-ending supply of energy. It's a deeply tempting prospect, one on the forefront of many a scientist's mind these days.

Just as Jupiter outsizes Earth, many other stars dwarf our sun. A spectrographic analysis of the heavens reveals stars that have surface temperatures 50,000 times greater than the sun, and cores as high as 2 billion K. In theory, stars could get even larger and hotter, but these nuclear furnaces create something more than just heat and light, something very tiny but that can limit the temperature a star may reach: neutrinos.

Neutrinos are so unbelievably small and slippery that they can travel at near the speed of light through virtually anything. Each second, the sun releases about 2×10^{38} of these little guys (that's 200 trillion trillion trillion neutrinos), and about 65 billion of them end up passing through each and every square centimeter of Earth. That's trillions of neutrinos passing through you right now. It doesn't matter if it's nighttime and the sun is shining on the other side of the planet; neutrinos can literally cruise through Earth, meandering among the atoms, unaffected and unaffecting.

Neutrinos are small, but they do carry a little bit of energy away from a star, and as a plasma reaches about 4 billion K, the atoms become so energetic that the neutrino production actually begins to cool the star significantly. However, if a star becomes massive enough, and hot enough to reach about 6 billion K, the heat triggers such a massive release of neutrinos that the star collapses and then explodes into a colossal supernova. So 6 billion K effectively sets the maximum temperature for a star.

During a supernova, though, things can get crazy and all bets are off on the temperature scale. In 1987, astronomers witnessed a supernova in the Large Magellanic Cloud (one of only two galaxies close enough for us to see with our unaided eye). By careful analysis, they determined that the temperature inside the explosion reached about 200 billion K.

So was that the hottest thing in the universe? Far from it. In fact, you can find a hotter spot just an airplane ride away.

Like obsessive psychoanalysts, many physics researchers insist that the only way to truly understand our universe is to look back into its infancy, to see what crazy things happened within the first second after birth. Back then the heat must have been unbelievably greater than a puny supernova, somewhere in the trillions of degrees. At that temperature, atoms should not only shed their electrons, not only break into their constituent protons and neutrons, but also literally melt into a plasma of quarks and gluons—a seething, primordial broth of elementary particles.

With that in mind, physicists at the Relativistic Heavy Ion Collider at the Brookhaven National Laboratory on Long Island, New York, have been accelerating heavy gold ions around a huge underground ring, speeding the ions up to 99.99 percent of the speed of light, and smashing them into one another. The result is an enormous quantity of heat in a very small space, and in 2010 they achieved a record-setting 4 trillion K (over 7 trillion °Fahrenheit). The experiment confirmed the science, creating a quark-gluon plasma. However, the scientists were surprised to find that the result was more like a "quark soup" than a gas; subsequent calculations indicated that a million times more heat would likely be required to boil it.

On the outskirts of Geneva, Switzerland, CERN's Large Hadron Collider is currently attempting to do just that. Scientists have already achieved trillions of degrees by smashing heavy lead atoms together; how long until the quadrillion- or quintillion-degree mark is smashed, too?

There is, nevertheless, a theoretical upper limit on the thermometer. We know that the hotter particles get, the faster they move. But Einstein also figured out that as particles approach the speed of light, they also increase in mass. Keep increasing the temperature, and at some point each particle of matter would become so dense that it would collapse into its own black hole, causing a minor disruption in . . . well, pretty much everything. The German physicist Max Planck calculated that this would happen at about 1.4×10^{32} K. That's 140 million million million million million degrees. And that, at least in this universe, is absolute hot.

The Creator and Destroyer It is no surprise and no coincidence that virtually every religious tradition describes a divine life-giving warmth that cleanses and sanctifies—but that can also punish or annihilate. For heat is the creator, and every atom in your body and

▲ Max Planck

beyond was fused in the fiery depths of a star, often at the moment of its own supernova. And just as surely, heat is the destroyer, rending the elements apart, ending one form and transmuting it into another.

Heat is also the mover and shaker, allowing energy to radiate, infuse, and enable reactions throughout the universe. Without it, molecules could not have bonded together to form the amino acids and other fundamental structures that led to the spark of life, nor could the myriad chemical reactions required to sustain that life—your life—endure.

And yet, watching the ignition of Trinity, the first test of an atomic bomb in 1945—the heat from which melted the New Mexico desert into a crater of radioactive glass 300 meters (1,000 ft) wide—the physicist Robert Oppenheimer recalled lines from the holy Hindu scripture the Bhagavad Gita:

> If the radiance of a thousand suns
> Were to burst at once into the sky,
> That would be like the splendor of the Mighty One...
> I am Death the destroyer of worlds.
>
> —*Bhagavad Gita, chapter 11:12, 32*

Agni, the Hindu god of fire from the 3,500-year-old Rig Veda scriptures, represents the essential life force in the universe, the creator of the sun and stars and the receiver of burned sacrifices, consuming and purifying so that other things may live. Agni may have also given birth to something else: the Latin word *ignis* ("fire"), which begat our English words *ignite* and *igneous* ("having formed from lava").

Nevertheless, heat offers hope, known by any hiker in the wilderness waking to a new sunrise. But removing heat, cooling elements, also offers hope: relief from a scorching daylight, or—in the laboratory—the glimmer of a possibility that we will better understand the building blocks from which we are all made. Cold creates order, though the *very* cold appears to create new disorders that we're just beginning to comprehend.

We are all phoenixes, born from the ashes, living in the radiant glow of the sun—"the force," as Dylan Thomas wrote, "that through the green fuse drives the flower." Heat is the spectrum of life, however you measure it.

TIME

Time keeps on slippin' . . . into the future.

—Steve Miller

THE GREATEST MYSTERY OF OUR UNIVERSE IS A PHENOMENON SO
precious it is considered by some to be holy, but so common that
it is for the most part ignored. More elusive than the slipperiest
substance, yet completely unavoidable, it is the enigma of time. Even
a toddler understands the passing of time, but the most insightful
scientists and philosophers remain baffled as to what it is, how it
works, and how best to measure our place in its ephemeral and
inexorable rhythms.

Of course, a glance at your wristwatch will tell you everything
you probably need to know about time right now, but what does the
ticking of each second really indicate? In the time it takes for you
to read this sentence, our planet will travel 300 kilometers (185 mi)
around the sun, 42 babies will be born, and your laptop computer
could calculate 40 million different chess moves. So what do we
make of the only slightly larger handful of seconds we call our
lifetimes? Truly, in order to understand time, you must let go of your
expectations of what *short* and *long* mean. After all, a millionth of a
second is a long time for a subatomic particle, and a million years is
just a blink of the eye from a cosmic perspective.

The only thing we can definitively say about time is that it
involves change of some sort—without change, there is no time. As
the Greek philosopher Heraclitus wrote, "You cannot step twice into
the same stream." (This is often translated more poetically as "No

man steps into the same river twice, for it is not the same river and he is not the same man.")

When talking about time, it's important to get clear what aspect of time you're discussing. For example, *duration* ("It took one second") is the flip side of *speed* ("In that time, the bullet traveled 300 meters"). But both of those are different from assigning a name to a moment ("At the tone, the time will be . . . noon"). To be sure, discussing time is tricky, but nevertheless an understanding of our universe requires an inspection and a measurement of time.

Measuring Time Sometime in the last ten thousand years or so, we caught on to the regularity of three natural cycles: the sun rises each day, the moon cycles each lunar month, and the sun returns to the same location each year. These simple events, experienced by all of us on Earth, lay the groundwork for the markings on every clockface and calendar.

However, you can imagine the confusion and frustration when early astronomers discovered that the year did not divide by an even number of lunar cycles, or lunar months by days! Instead, each lunar month lasts about 29½ days, and 12 of these lunar months span only 354 of the approximately 365¼ days in a solar year. Trying to make sense of these irregular numbers tried the patience of even the most devout timekeeper.

Nevertheless, the simple 12-month lunar calendar was expedient for the early Sumerian and Babylonian agricultural civilizations, and to this day it forms the basis of the Islamic religious calendar. (That is why the same Islamic holidays show up at different times each year; Islamic dates shift about 11 or 12 days each solar year, returning to the same position about every 33 years.)

A number of other groups, including the Jews and early Greeks, decided to reconcile the solar and lunar calendars, creating a complex and somewhat syncopated system involving 12 years of 12 lunar months interspersed with 7 years of 13 lunar months. On the

> **π times 10^7 seconds is a** good approximation of one year (it's 363.6 days). Even better is taking the square root of 10, then multiplying by 10 million seconds (almost exactly 366 days). But π x 10^{16} seconds is about 1 billion years (an eon).

Jewish calendar, every 2 or 3 years there's an extra month, called Adar I, thrown in.

The early Christian church divested itself entirely of the lunar calendar, creating first the Julian calendar and then the slightly more accurate Gregorian calendar, which most of the world uses today. Some say the move was a conspiracy against the feminine aspect of spirituality (as the moon is considered a symbol of women's power); others point out a more practical reason for the church's decision: The farther from the equator you are, the more important the solar seasons are relative to the lunar cycles (which you may not even be able to spot through the cloud cover).

Curiously, the ancient Egyptians had a slightly different compromise to the calendar problem: They rounded the "month" up to 30 days. Twelve of these equal 360 days, a number particularly propitious to mathematicians of the time, who had standardized on the ancient Sumerian sexagesimal (base-60) counting system. After all, if you're primarily concerned with dividing parcels of land, or goods at market, or times of the year, counting by 60 is extremely efficient because it can be divided equally by so many numbers: halves, thirds, quarters, fifths, sixths, tenths, twelfths, and so on. Sixty, and by extension 360, is a beautiful number when you don't have a calculator. In fact, there are still communities of people in Asia who count on their fingers by pointing with their thumb to each finger bone on one hand (up to 12) and then keeping track of sets with fingers of their opposite hand (up to 5), totaling 60.

So perhaps it was natural that if a year were split into 12, then the day and night would also each be split into twelfths, leading to 24 hours from one sunrise to the next. Each hour could then easily be split into 60 minutes and each minute split into 60 seconds, leading to a wonderful symmetry of 360 seconds per hour, like the 360 days per year. It all seemed so perfect, until the Egyptians recognized that the year actually required five or six extra days. It appears they reluctantly snuck these into their calendar, almost as an afterthought.

You know that each four years is a leap year and you have to add a leap day, February 29. And you may know that you have to omit that extra day every 100 years. But do you know that you have to put it back in every 400 years? That's why 2000 was a leap year but 1900 was not. And don't forget to remove the leap day every 4,000 years, or reintroduce it in each century year that, when divided by 900, leaves a remainder of 200 or 600.

Q: In what year was Friday, October 15, the day after Thursday, October 4?

A: In 1582, when Pope Gregory instituted a new calendar!

Although the days, months, and years all follow (more or less) astronomical intervals, the seven-day week is perhaps the most cosmic of all, as it's the only measurement that stems entirely from the Hebrew Bible. Exodus 20:9–10 states clearly that "six days shalt thou labor, and do all thy work, but the seventh day is the Sabbath of the Lord thy God." However, other than divine instruction, there's nothing particularly special about a seven-day week, and a number of other cultures standardized on other convenient weekly patterns of four, eight, or even ten days.

A 10-day week has long suited those attracted to the idea of "metric time," where each of the 12 months would be split into 3 weeks of 10 days, each day would contain 10 hours, and each hour would have 100 minutes of 100 seconds each. It sounds preposterous, but of course it's no more arbitrary than the clocks we use today. Metric time gained followers in the French Revolution's flush of reinvention, but the idea—with its deciday, milliday, and microday measurements—soon died out, in part because early nineteenth-century Christians felt its lack of a Sabbath proved the metric system was an abomination.

Since then, rather than reinvent the counting system, scientists have decided to declare ever more precise definitions of time measurements. In 1954, the General Conference on Weights and Measures decided that 1 second should equal exactly 1/86,400 of a mean solar day. Feeling that this definition was still too wishy-washy—after all, the length of a solar day literally shifts during massive geologic events—scientists searched the physical world for a universal standard and alighted on a number that should be reproducible anywhere in the universe: the frequency of microwave light that a heated cesium 133 atom absorbs or emits—specifically, the time it takes for 9,192,631,770 of these electromagnetic wavelengths to pass. True, this definition of a second, which was agreed upon at the 1967 conference, is a value only a scientist could love, but it set the stage for extraordinarily precise measurements of time.

▲ The FOCS1 caesium fountain atomic clock at the Federal Office of Metrology (METAS) in Switzerland

Today, the world's most accurate clocks are white-coat operations, cobbling together lasers, near absolute zero temperatures, and magnetic traps to manipulate the spin on electrons. Physicists on the cutting edge of science can now create timepieces based on quantum logic, accurate to within one second every 3.7 billion years.

This level of precision may seem like overkill to anyone who is just trying to catch a train, but a surprising amount of our science is based on the careful measurement of time. Remember that our basic division of space—the meter—is defined by how far light travels in 1/299,792,458 of a *second*. Similarly, both the fundamental scientific values of the lumen (which measures the amount of visible light from a source, like a lightbulb) and the ampere (or amp, which measures electrical current) are based on the value of a single second of time. Even measuring pressure in the real world relies on measuring force, and force is based on acceleration, which is determined by the passage of time! Without extraordinary clocks, the finest measurements would be impossible.

While scientists focus on measuring ever finer durations, the rest of us concern ourselves with the naming of time—assigning dates to historical events or setting our clocks based on political whims. It's important to remember that every chrononym (a word that refers to some particular time, such as "springtime" or "teatime") is arbitrary; every calendar is based on cultural norms and sometimes curious assumptions. According to Jewish scholars from the Middle Ages, our universe was created in 3761 BCE, which explains why the new millennium was celebrated as 5761 in some quarters. James Ussher, a seventeenth-century Irish Anglican bishop, calculated it slightly differently, noting that creation began the night before Sunday, October 23, 4004 BCE.

Obviously, the Chinese calendar is different from either of these, as are the Islamic and Hindu calendars. And many people put their faith in the Maya Long Count calendar, which prescribes that the fourteenth *b'ak'tun* (numbered 13.0.0.0.0) was to commence on December 21, 2012, signifying either the end of the world or at least the end of some people insisting it'll be the end of the world.

The capricious nature of calendars is equaled only by the somewhat random method of setting our watches. Only 150 years ago, clocks around the world were synchronized at noon by some local official's judgment on when the sun was directly overhead. For instance, in the early nineteenth century, the *Chicago Tribune* regularly reported the variances among fifty-four different local times across Illinois and Michigan alone. This was manageable when you worked almost entirely with people in your own town, but it quickly became untenable as people began to travel. The quick and vast expansion of the railroad, more than anything, created the need for a widespread sense of "official time," and it was the railroad companies that first instituted, in 1883, standard time zones across the United States and Canada.

> "O God! methinks it were a happy life . . .
>
> To carve out dials quaintly, point by point,
>
> Thereby to see the minutes how they run—
>
> How many makes the hour full complete,
>
> How many hours brings about the day,
>
> How many days will finish up the year,
>
> How many years a mortal man may live."
>
> —Shakespeare, *Henry VI, Part III*

The following year, at the International Meridian Conference, delegates from twenty-five nations agreed on a worldwide system that outlined 24 fifteen-degree wedges around the globe, each marking an additional hour forward or backward in time. The "zero point" from which all zones are measured is in Greenwich, England—leading to Greenwich mean time (GMT).

In a perfect world, these smooth lines would run from pole to pole like stripes on a beach ball. But a wide spectrum of commercial and geopolitical interests has made these lines ridiculously ragged. The most extreme example is China, which until 1949 contained five different time zones but now maintains one "Beijing time" across the entire nation—even though this means "noon" doesn't come until the middle of the afternoon for cities in western China.

The establishment of GMT also created another arbitrary demarcation: the international date line, which runs more or less through the middle of the Pacific Ocean. When you travel east across

▼ The 24 world time zones rarely follow straight lines. Stripes indicate regions that are ½ hour different.

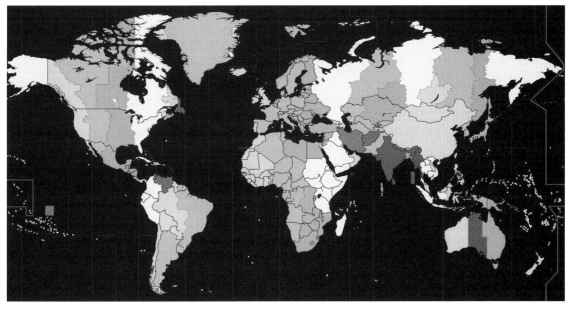

CIA World Factbook

the date line, you jump forward an hour* but back a day. A traveler from Tokyo to Seattle, therefore, experiences the odd sensation of technically arriving before she left.

Here's another example of human-created time line oddities: In 1995, the tiny country of Kiribati, comprising more than thirty small atolls and islands that straddled the international date line, decided to redraw the line. (Some have cynically suggested this was an attempt to increase tourism, as it would suddenly jump to the head of the line as the first country to enter the twenty-first century.) The change means that today, while eastern Kiribati and Hawaii have approximately the same longitude, calendars in Kiribati are bizarrely a day ahead *and* two hours behind Honolulu. Samoa, not content to slouch behind, also recently jumped ship, deciding to entirely skip December 29, 2011, and leap into the future a day.

Imagine an alien arriving on Earth, attempting to make sense of the mess we've made of time. From seconds to light-years, our measurements can be understood only from our provincial, terrestrial perspective. And yet, if we were to try to explain our sense of time to that visitor, we would have to start at the very beginning: the durations and speeds that we as humans can actually sense and comprehend.

Human Time A second, an hour, a year—these are all part of human time, time we recognize, time at the scale in which we live. We have named each aspect of our human time, sometimes with language as flowery as a perfume commercial. It's happy hour, or the witching hour; it's as slow as molasses, or as fast as a bat out of hell. The fastest events happen in just a blink of an eye, or in a flash. That makes sense, as a blink of an eye takes about a third of a second, and our eyes can reliably see events down to about a tenth of a second. Anything faster than that becomes a blur, or doesn't register at

---

*Actually, in some areas, the date line and the time zone lines diverge, so this isn't always technically accurate.

all. But project a flipbook of static images in succession at twenty or thirty frames each second and they blend together seamlessly, fooling our eyes into believing we see smooth movement. Thus the movie was born.

On the other end of our human sense of time, a very rare event—one with a long duration between occurrences—happens perhaps once in a generation or once in a lifetime. A generation is usually considered twenty to twenty-five years; a lifetime, only a few times that. Scientists have discovered that most animals live, on average, only about 1 billion to 1.5 billion heartbeats. This is why larger animals, with slower and more efficient metabolisms and heart rates, typically live longer than small, fast-heart-beating creatures. There are exceptions, of course. Parrots can live as long as a human, though they have twice our heart rate. And humans have gamed the system with medical advances, so that we now average about 2.5 billion heartbeats.

But who is really living those billion-plus heartbeats? As the roboticist Steve Grand has pointed out, we are not who we were as children, and our earliest memories could almost be ascribed to someone else. After all, our entire blood supply is re-created anew every three months, our skin replaced every two weeks; in fact, every cell in our body is replaced at least once each decade, and there are few molecules in us today that were with us as children! Our sense of self remains consistent throughout our lives, as do our memories, but in truth we are each like a mountain lake, constantly renewing as we empty, the flow maintaining our form (more or less) and warding off stagnation.

Nevertheless, the thing we call our self is living for around a century, give or take. From one perspective, a lot can happen during those approximately three billion seconds; but from another perspective, very little.

A single second—the building block on which all other time is now measured—is the time it takes an average heart to beat once, or the time in which we pass 1 meter (about 3 ft) during a leisurely

> "Which is quicker, a jiffy or a flash? I think there are two flashes in a jiffy, myself. But God knows how many jiffies there are in two shakes of a lamb's tail. But why did they use two shakes of a lamb's tail? What's wrong with the basic unit of measurement, one shake of a lamb's tail? We can do our own arithmetic, thank you."
>
> —George Carlin

In the early years of train travel, passengers propelled up to 55 km/h (35 mph) described the experience as "breathtaking."

stroll. But in that one second, a hummingbird flaps its wings 70 times, sound travels 340 meters, and a flash of light rushes past 300 million meters.

A 90 mph fastball takes only half a second to reach home plate, and a batter must decide whether to swing in just a portion of that time. It took Jamaican-born Usain "Lightning" Bolt only 9.58 seconds to run 100 meters in 2009—an unheard-of act, an average of about 10 meters per second (that's 38 km/h or 23 mph).

Of course, these human acts pale in comparison to abilities in the animal world, where the cheetah and sailfish tie the speed records, one running and one swimming, covering 31 meters in a second (113 km/h or 70 mph). The peregrine falcon beats them all, but only with the help of gravity's pull: Nested atop cliffs (or, increasingly, on skyscrapers in large cities), it can dive toward food at more than 90 meters per second (322 km/h or 200 mph).

At a scale too small to detect, oxygen molecules at room temperature zip through the air at over 450 meters per second (1,600 km/h or 1,000 mph). And at a scale too large for us to sense, Earth spins through space at about the same rate—the equator turning at about 1,680 km/h. An aircraft or bullet passing through air that fast would create a sonic boom, but in this case, Earth literally drags the air with it, so we're spared the noise.

Not all planets turn so rapidly; Venus, for example, rotates only 1.8 meters each second (4 mph)—at this rate, you could walk its equator at a brisk pace and appear to stop time, or at least stop the sun's progress across the sky.

That said, most objects in space tend to stretch the limit of our ability to understand speed. In a single second, a communication satellite, traversing the sky 36,000 km (22,000 mi) above sea level in a round-each-day geosynchronous orbit, travels 3,100 meters in a second (11,160 km/h or 6,900 mph). The space shuttle was twice as fast, accelerating from 0 to 17,000 mph (27,360 km/h) in just over 8 minutes. But that's only fast enough to maintain an orbit! To propel a rocket beyond Earth's immense gravitational pull, it must reach speeds of over 11,200 meters per second (40,320 km/h or 25,000 mph). That still isn't fast enough to escape the sun's orbit and break away from our solar system—a feat that requires an object to travel almost four times faster.

At the opposite end of our human scale of time, a sloth, infamously slow due to a hypervegetarian diet lacking in protein and fats, can travel only about 10 cm per second (0.2 mph); a snail, limited by its mode of transport as much as its size, slimes along a tenth as fast as that. Although they may strain our patience, we can see these movements, which is not the case for many other phenomena around us. While it varies widely from person to person, hair grows, on average, about 4.5 nanometers per second—that's impossible to see on short time scales but easy to perceive as stubble growing at ⅜ millimeter per day, or about 1.25 centimeters (0.5 inch) per month. You might call these speeds "glacial," but hair grows far slower than glaciers migrate—as quickly as two tenths of a millimeter per second (17 meters or 56 feet per day).

Nevertheless, the speed of a glacier, or hair—or even a flower sprouting, blooming, fading, and dying—are all well within our range of understanding. We can wrap our heads around these human scales in a way that we cannot fathom as we step beyond the century and millennium, or look more closely at events faster than a flash.

Geologic Time Our planet is about 4.5 billion years old, originally formed from floating interstellar rock and ice pulled together by gravity. Imagine that a single year is represented by a 1-millimeter length of string; a century is 10 centimeters (about 4 in.), a millennium is 1 meter. To demonstrate the age of Earth, you would need a string that spans from San Francisco to New York.

Obviously, for those of us who usually consider "long term" as having to do with our retirement, it takes a while to warm up to a perspective from which "soon" means sometime within 20,000 years. After all, 20,000 years compared with the age of our planet is like one minute in five months, or three hours within a human lifetime.

In thoroughbred horse racing, one length is about one fifth of a second.

The oldest living individual organism on Earth is a Great Basin bristlecone pine tree (*Pinus longaeva*) in the mountains between California and Nevada. It is 4,842 years old, based on a ring count of a sample core. This tree, known as Methuselah, was more than two centuries old when the Great Pyramid of Giza was constructed. If you include plants that clone themselves, the longest-living organism may be a single clonal colony of *Populus tremuloides* (quaking aspen) in Utah that has been growing for an estimated 80,000 years.

Here's another way to think about it: You've likely seen time-lapse movies of clouds moving or even seasons changing, but imagine a time-lapse where each frame captures a moment every ten thousand years. If we started shooting at the birth of the planet, the finished movie (to the current day, at least) would be four hours long, and the entire history of the human species wouldn't show up until well into the final one second of footage.

At this rate, what scientists call geologic time, our sense of how the world works begins to break down. For example, over thousands of years, diffusion—the process in which molecules comingle, like gas mixing in a container—affects solids. That's why gold and lead objects found next to each other in Egyptian tombs have merged, as though they have melted together.

On a slightly longer scale, there is no doubt that humans have significantly affected our planet's climate in recent years, but climatologists must also take into account the normal weather cycles, which can be seen only by taking a geologic view. For example, Earth's axis (which points roughly toward the North Star) rotates slightly but returns to the same point every 25,784 years—about as long as it takes a ray of light from the center of our galaxy to reach us on Earth. Think about it: The last time the stars were in the same location they are today, humans were painting in caves during the last ice age.

This normal wobbling is added to a 41,000-year cycle where the tilt of the poles varies between 22.1° and 24.5° and another cycle in which every 100,000 years Earth's orbit shifts from elliptical to circular, then back again. It doesn't sound like much, but a few degrees mean the difference between a lovely spot of weather and an ice age that blankets much of the planet in snow. The irony of our current global-warming problem is that, because the axis tilt is decreasing, we may be heading toward another natural ice age in ten or twenty thousand years, and the warming of the planet may push it back a bit.

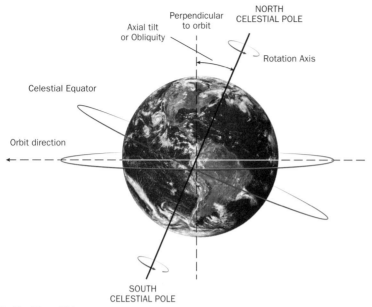

NORTH
CELESTIAL POLE

Perpendicular
to orbit

Axial tilt
or Obliquity

Rotation Axis

Celestial Equator

Orbit direction

SOUTH
CELESTIAL POLE

▲ Earth's wobble

Extending our perspective over an even larger span of time, we know that landmasses on floating tectonic plates move at about the speed fingernails grow—South America and Africa are moving apart and the Atlantic Ocean is widening at about 4 centimeters per year. So over the next million years, Los Angeles will creep about 40 kilometers north-northwest of its present location. Even a million years—called a megannum (abbreviated Ma)—isn't that long: If we could build a spaceship that traveled at the speed of light, a million years wouldn't get us to even the halfway point on a journey to the Andromeda galaxy.

Scientists who think in terms of megannums like to mark off the ages of Earth's development, like lines on a doorframe as a child grows. In case you're taking notes, at the time of this writing, we're in the Phanerozoic geologic eon (which started about 542 Ma

Earth's rotation is slowing, due to the gravitational pull of the moon on the tides, leading the length of the day to increase by about 2 seconds every 100,000 years.

ago), in the Cenozoic era (which began when the dinosaurs died off, about 65 Ma ago), and—for the last 11,000 years—the Holocene epoch. Some argue we're also in a new era, called the Anthropocene, characterized by a vast array of human-created sediments spread about the planet's surface (plastic and other refuse, mostly).

It's odd to think about, but sediment and waste—what animals and plants create during their short lives and leave behind afterward—actually make up a huge portion of Earth's outer layers, from rock we walk on to the atmosphere we breathe. For hundreds of millions of years after the creation of Earth, the planet was a superheated ball of sterilized rock and gas, bombarded with asteroids, devoid of any life. But about 3.5 billion years ago, it cooled down enough to support shallow pools of water, in which complex molecules combined to form tiny single-celled bacteria. The bacteria lived, divided, and died, leaving behind their microscopically small shells. A few trillion would have made no difference, but trillions of trillions, over hundreds of millions of years . . . well, it all adds up, and the sediment slowly built up in layered mats that today can be seen as stromatolite rock.

Then, about 2.5 billion years ago, as available organic food sources became scarce, the bacteria learned a new trick: Take water, carbon dioxide, and sunlight and convert them into energy. Unfortunately, this clever technique, called photosynthesis, releases a deadly waste product: oxygen. Well, it was deadly to them at the time, just as carbon dioxide (which we exhale) is toxic to us.

Most of the oxygen combined with iron and other molecules, which over time got trapped in thick layers of sediment (what we today mine as iron ore). Eventually, though, about a hundred million years later, all the available iron oxidized and the oxygen had nowhere to go but up into the atmosphere, resulting in the Great Oxygen Catastrophe, during which the majority of life on Earth died. Of course, one person's toxic waste is another person's chance at life, and it wasn't long (remember, we're talking geologic time here) before tiny single-cell and then multicellular creatures appeared

> **"Time is the fire in which we burn."**
> —Delmore Schwartz, "Calmly We Walk Through This April's Day"

A lustrum is 5 years (from the Latin "to wash," as a form of purification after the census was taken every 5 years in Ancient Rome).

that could take advantage of the oxygen-rich air and the brand-new protective layer of high-altitude ozone (a molecule made of three oxygen atoms that stops most harmful radiation from getting through from space).

There is an ongoing debate between creationists and evolutionists, in which the former insist that complex structures (such as the eyeball) could not have evolved—there are too many different but interrelated parts that must work in perfect synchrony. But what most creationists don't take into account is the incredibly long periods of time involved. Of course an eye could not develop in a year, or even a million years. Evolution of plant and animal life—like sediment—happened over a time span that we, as humans, can scarcely comprehend. It took perhaps 400 million years for multicellular organisms to evolve into the simplest animals, like sea sponges or jellyfish; another hundred million years for fish; then insects and reptiles showed up over the next couple of hundred million years. Throughout this time, plants grew huge, developed, died, and—along with the animals—left immense amounts of rich organic sediment that we now know as fossil fuels, such as coal, oil, and natural gas.

That all happened long before dinosaurs appeared, about 240 million years ago. The dinosaurs ruled Earth for a measly 175 million years, then were suddenly wiped out, probably within hours of a six-mile-wide meteor hitting near the eastern coast of what is now Mexico. Once again, when one door closes, another opens, and the animals that survived (the ones that could burrow, swim, and adapt quickly) had a tabula rasa with which to start anew.

Finally, not until about two and a half million years ago—63 million years after the great dinosaur extinction—did the earliest forms of genus *Homo* appear, and not until about 100,000 years ago did our particular species of *Homo sapiens* show up.

Once again, these geologic time spans are so vast that in order to find our place in them, perhaps we should try to compare them with something we know well: our own human life span. Let's imagine

> **"Time is the substance from which I am made. Time is a river which carries me along, but I am the river; it is a tiger that devours me, but I am the tiger; it is a fire that consumes me, but I am the fire."**
>
> —Jorge Luis Borges, "A New Refutation of Time"

that "Mother Earth" is celebrating her 50th birthday right now—that is, her whole 4.5 billion years were compressed down into 50. In that case, the first signs of simple life didn't show up until she was age 11. The first animals with eyes (such as trilobites and horseshoe crabs) appeared just before her 45th birthday. Dinosaurs went extinct a few months after she turned 49. *Homo erectus* (Java man) learned to control fire this week, and *Homo sapiens* showed up yesterday. The last major ice age ended an hour ago, and civilization—virtually all our recorded history, art, and science—began a few minutes later. Jesus was born a little over 10 minutes ago. And the age of modern computers began 17 seconds ago. Happy birthday, Mom.

Cosmic Time Once you've trained yourself to think in terms of billions of years—assuming that's even possible—you can see patterns of movement and change at a galactic and universal level that is impossible to consider by simply looking at the sky on any given night. That's not to say that astronomical objects are moving slowly! They're just moving slowly compared with much larger things; it's all relative. Within our solar system, for example, the fastest-moving planet is Mercury, named after the Roman messenger of the gods, which orbits the sun at about 48,000 meters per second (about 107,000 mph); Earth is lazy in comparison, managing only about 30 km/s (67,000 mph).

The sun, and our solar system as a whole, is traveling through space, rotating around the center of the Milky Way galaxy at 800,000 km/h (500,000 mph). And while all that is happening, our galaxy is moving through space at over 2 million km/h (1.2 million mph). That all sounds breakneck until you consider that it takes more than 225 million years for us to make a single orbit—one galactic year.

Our sun is about 4.63 billion years old, or about 20 of these galactic rotations. Back when it was created, five eons ago (an eon is 1 billion years), we were all just a nebula of gas, ice grains, and dust floating through space—mostly carbon, oxygen, silicon, and iron generated in the dramatic explosions of earlier stars. Some

> "Science cannot solve the ultimate mystery of nature. And it is because, in the last analysis, we ourselves are part of the mystery we are trying to solve."
>
> —Max Planck, physicist

shock wave from a nearby supernova may have shoved enough of this matter together that it condensed into a star and planets, similar to how dust bunnies clump under your bed. The universe is like that: Stars grow from dust, then live and die, and sometimes blow up, offering life to new stars and planets. It's hardly newsworthy; billions of stars are born and die each year across the vast reaches of our universe.

It's likely that genesis—what some folks call the big bang— happened about 13.7 billion years ago, plus or minus 110 million years. That's almost 14 million millennia—though only about 60 of our galactic years. Consider Carl Sagan's famous "cosmic calendar," in which the history of the universe is compacted into a single solar year. At this scale, if the universe began on January 1, the Milky Way started forming in March, though our sun and Earth weren't born until September 1. Later that month, single-celled creatures appeared, followed sometime in November by multicellular organisms. Mammals appeared on December 26. Dinosaurs were wiped out on December 29. All human prehistory (from the first known stone tools) and history occurred during the final hour of New Year's Eve, and our entire recorded history of civilization takes up just the last 22 seconds of the last minute of December 31.

The amazing thing is that the universe is still relatively young and impetuous, changing all the time. Only a billion years from now, the sun will likely grow so hot that all life is extinguished from Earth's surface. In a few billion years, there's a small chance that fluctuations in the planets' orbits will cause Earth to crash into either Mars or Mercury, and a very good chance that the Andromeda galaxy will smash into ours, forming one incredibly big elliptical galaxy. Our sun will probably survive this upheaval but go on to burn out seven billion years from now. (There's an old science joke about the woman at the back of the lecture hall who, upon hearing this news, nervously asks for a clarification, then says, "Oh, *billions*! Goodness me, I was worried; I thought you said *millions*.")

If you compressed the age of the universe into a single day, the average human life span would take only 1/2,000 of a second.

While there is nothing older than the universe (at least nothing we can measure), scientists often calculate far longer lengths of time, particularly when considering astronomical lifetimes. The estimated life span of a red dwarf star about a tenth the mass of our sun is 312 exaseconds—that's 10^{18} seconds, or roughly 10 trillion years. With that in mind, given our current understanding, the last star in the universe will likely die about 100 trillion years from now. Nevertheless, the universe itself will go on, primarily made up of black holes, which evaporate incredibly slowly due to something called Hawking radiation. One relatively small black hole the mass of our own sun would take about 6.6×10^{50} yottaseconds to burn away—that's 6.6×10^{74} seconds, 2×10^{67} years, or over a trillion trillion trillion trillion times longer than the current age of our universe.

These numbers are staggering, and yet, just as the largest stars gain their size and energy from the smallest atoms, even the longest events are constructed of the thinnest slices of time, one moment after another.

Very Fast Things In his book *Metamagical Themas*, Douglas Hofstadter describes how, in the 1940s, Nicholas Fattu led a team of ten people working full-time for ten months to solve a tricky math problem. Twenty years later, he used a room-sized mainframe computer to find the solution in just twenty minutes (which included discovering errors in the original work). Today the solution could be found even more accurately in less than a second using a laptop computer. But what's really going on in that single second? What would we find if we slowed time, expanding it until we could peek in on events that appear virtually instantaneous?

You can see individual images flickering by at a rate of ten per second; you can hear individual clicks or strums that fast, too. But speed up the film or soundtrack and a magical thing happens: Our brain blends the experiences together, creating an amalgam, a new experience different from the sum of its parts. Within a tenth of a second, a small hummingbird can beat its wings 7 times, creating

Speed

Fingernails grow	1.2×10^{-9} m/s (2.6×10^{-9} mph)
Moon receding from Earth	1.3×10^{-9} m/s (about 1 nm/s)
Average growth rate of child	1.8×10^{-9} m/s (4×10^{-9} mph)
Growth rate of bamboo	6×10^{-7} m/s (1.3×10^{-6} mph)
Garden snail	0.002 m/s (0.004 mph)
Audiocassette tape speed	0.0476 m/s (0.106 mph)
One knot (nautical mile per hour)	1.852 km/h (1.151 mph)
Average walking speed	1.2 m/s (2.5 mph)
World record swimming	2.3 m/s (5.2 mph)
Comfortable bicycling speed	6 m/s (13 mph)
Fastest human running	12 m/s (27 mph)
Horse in gallop	13 m/s (30 mph)
Object falling 10 m	14 m/s (31 mph)
Sky-diver in midflight	54 m/s (120 mph)
Sneeze	up to 46 m/s (104 mph)
Golf ball driven off tee, arrow fired from longbow	60 m/s (130 mph)
Stock car	90 m/s (200 mph)
Wind in tornado	112 m/s (250 mph)
Fastest train with wheels (non-maglev)	160 m/s (357 mph)
.22 long-range centerfire bullet	400 m/s (1,300 ft/sec, or 895 mph)
Earth's rotation at equator	464 m/s (1,038 mph)
SR-71 Blackbird, fastest jet aircraft	981 m/s (2,194 mph)
North American X-15 rocket-powered aircraft	2,020 m/s (4,519 mph)
Satellite in geosynchronous orbit	3,100 m/s (6,900 mph)
Space shuttle, and International Space Station in orbit	7,743 m/s (17,320 mph)
Apollo 10 manned spacecraft, return trip from moon	11 km/s (24,791 mph)
Earth orbits the sun	30 km/s (66,620 mph)
Helios space probe (orbiting the sun)	70.22 km/s (157,078 mph)
Solar system orbiting the Milky Way	216 km/s (483,000 mph)
Milky Way moving through space (relative to cosmic background radiation)	550 km/s (1.2 million mph)
Rotation of fast neutron star	38 Mm/s
Light in a vacuum	299,792,458 m/s (186,282 mi/sec)

0 20 40 60 80 100 120

MPH

a low hum in our ears. Vibrate a string 44 times in the same tenth of a second, and our ears hear A above middle C. In that same split second, the sound will travel 33 meters (108 ft) through the air, but a light flipped on at the same moment travels quite a bit farther—in a tenth of a second, the light could travel three quarters of the way around Earth. If that light wave were an AM radio signal at the lowest end of the dial, it would have vibrated 55,000 times in that flash of light.

It takes two tenths of a second to react to something you see— slightly less for an auditory stimulus—but it takes just a hundredth of a second for a nerve impulse to travel from your brain to your hand. Actually, signals travel at varying rates through different neurons (nerve cells), from 1.6 to 600 km/s (1 to 268 mph), which sounds fast, but it's millions of times slower than electricity flowing through wire. When you touch something, electrochemical impulses send the message to the brain extremely quickly, but if that thing is sharp or hot, you won't recognize that until later, because pain signals travel 100 times slower.

We've become familiar with hundredths of seconds in large part due to the sports we watch. While human reflexes are not fast enough to judge even tenths of seconds, electronic instruments are clearly far more accurate. In the 2008 Olympics, when Michael Phelps won his seventh gold medal for swimming in the 100-meter butterfly, his time was 50.58 seconds, just one hundredth of a second faster than Milorad Čavić. The time was based on an ultrathin plastic touchpad mounted on the wall and was reinforced by high-speed cameras running at a hundred frames per second. Phelps was recorded slamming his hands down on the pad one frame faster than the Serbian.

The very fastest aspects of our lives can be captured at the scale of milliseconds (thousandths of a second). A housefly flaps its wings once every three milliseconds; a normal point-and-shoot camera can record a single photograph in good light, stopping human motion, in one millisecond. In running and bicycle racing, extremely fast

> **"Time is an illusion. Lunchtime doubly so."**
>
> —Douglas Adams, novelist

cameras, shooting one thousand frames per second, focus on the final moment at the finish line. Coaches and athletes will argue, but the difference between gold and silver at this scale—especially when it comes to swimming, running, bicycling, skiing, or horse racing, where the competitor's environment can change uncontrollably from one moment to the next—is a matter of luck as much as skill.

When most people think about financial trading, they imagine people yelling on a crowded market floor, but today the big trades are done not by humans but by silent computer algorithms making split-second decisions based on news collected from around the world. Case in point: A new transatlantic cable is being laid, at the cost of $300 million, in order to shave six milliseconds off the time it takes for a signal to travel from Europe to New York. Many analysts applaud this move, noting that even a single millisecond's advantage means an extra $100 million each year to a large hedge fund.

We can somehow understand milliseconds: A thousand events in a second is shocking but still within comprehension. However, microseconds (μs)—millionths of a second—teeter on the edge of disbelief. A microsecond is to a single second what 1 second is to 11.5 days—in other words, if you took one step each second, you'd have to walk for 277 hours straight to reach a million. To give you a sense of how long a microsecond is, it takes about 500,000 μs to click a mouse. And yet, because sound reaches one of our ears just 600 μs before the other, we can identify where it originated and turn our head to it, even with our eyes closed. (How we manage this is actually a mystery, as the signal from the ear to the brain may take as long as a millisecond.)

At these rates we also begin to encounter subatomic particles. For example, the entire life span of a muon—generated by powerful atomic collisions that transform energy to mass—is only about 2.2 microseconds, after which it quickly ruptures into an electron and two neutrinos. A lot can happen in even a single microsecond: Earth orbits another 18.5 millimeters (about ¾ in.), light travels 300 meters (1,000 ft), proteins in our cells stretch and fold into

The half-life of a neutron in isolation is about 10.5 minutes. That's a billion times longer than the half-life of any other known atomic particle.

complex three-dimensional shapes in order to carry out their life-enabling tasks.

What, then, of a nanosecond—a billionth of a second? Surely nothing could be that fast. And yet a microprocessor inside a desktop computer takes just a few nanoseconds to carry out an instruction, such as adding two numbers. In the time it takes you to blink your eye, a typical computer can do 900 million calculations. Suddenly Nicholas Fattu's experience starts to make sense: There are as many nanoseconds in a single second as there are seconds in 30 years—an entire lifetime of math performed by a human could be reproduced in a matter of seconds on your smartphone.

However, at the smallest measurable sizes, nanoseconds can seem like an eternity. Remember that when you see a red light, it's electromagnetically vibrating 400,000 times each nanosecond. The fastest switching transistors—the hardware that lies at the heart of a computer—operate in the realm of trillionths of a second, or picoseconds. Imagine living at this scale: Photons of light cover only one millimeter in a picosecond.

It takes the molecules in your eye only about a fifth of one picosecond—200 femtoseconds—to react to visible light. That light itself is constructed of wavelengths flip-flopping between electric and magnetic fields every two to four femtoseconds. These atomic speeds seem untouchable by humans, and yet today's fastest computers, which are measured in petaflops (quadrillions of operations each second), can complete a calculation in less than one femtosecond. At this rate, it's hard to tell a split second from a split-split-split second, but consider that there are as many femtoseconds in 1 second as there are seconds in 31.7 million years. Or, as *Scientific American* magazine put it, "More femtoseconds elapse in each second than there have been hours since the big bang."

In the late nineteenth century, Eadweard Muybridge stopped a racehorse in midrun by capturing a brief moment in time on a photographic plate, proving that all four legs were in the air at the same time. By the 1980s, scientists used the same techniques, but

The individual H$_2$O molecules in water have a slight attraction to each other, bonding for just a few picoseconds at a time. It's like a disco dance floor of constant movement, where molecules partner off, then break up and dance with others—the attractions keep it tight enough to form a mob but loose enough so that foreign molecules can easily wander through the liquid.

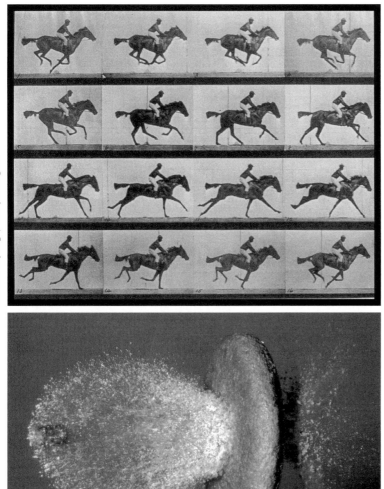

Eadweard Muybridge / Library of Congress

Jeff Krug

▲ Eadweard Muybridge's camera stopped a horse in 1881. Today's cameras are faster than a speeding bullet.

"For what is time? Who can easily and briefly explain it? Who even in thought can comprehend it, even to the pronouncing of a word concerning it? But what in speaking do we refer to more familiarly and knowingly than time? And certainly we understand when we speak of it; we understand also when we hear it spoken of by another. What, then, is time? If no one ask of me, I know; if I wish to explain to him who asks, I know not . . . How then can these two kinds of time, the past and the future, be, when the past no longer is, and the future yet does not be."

—Saint Augustine of Hippo,
Confessions

this time freezing never-before-seen molecular reactions by firing femtosecond laser pulses like a camera strobe light—flashes of light one millionth of a billionth of a second. It's an extraordinary accomplishment, but individual atoms still appear as nothing but a blur at this speed; an electron has completed its entire virtual orbit around an atom in that single femtosecond. Clearly, if we're going to travel into the impossibly small world inside the atom, we need to look at even smaller slices of time, at the very limits of light, size, and even matter itself.

Quantum Time As far as we can tell, everything in our universe is tied to the speed at which light can travel, and light travels really, really, really fast: 299,792,458 m/sec; or 186,282 miles, 698 yards, 2 feet, and just over 5 inches. Everyone agrees it's far easier simply to memorize "about 300,000 km/s."

Remember that light may travel far slower than this, depending on what it's traveling through: in water, it's about 75 percent as fast; through a diamond, it's only 40 percent. Scientists can actually force charged particles to travel faster than light through a dense medium, which creates a bizarre electromagnetic shock wave called Cherenkov radiation. But even then, the particles can't go faster than the maximum limit of light in a vacuum. Of course, that limit refers to travel only through the four-dimensional medium called space-time, and it's possible that light (or particles, such as neutrinos) may be able to travel faster if we someday discover additional dimensions.

Nevertheless, our understanding of the speed of light, along with our current grasp of the size of the universe, leads to some interesting realizations. Foremost, you can forget those fantastic images from *Star Trek* or *Star Wars* where the stars suddenly race past as you move beyond light speed. Even if you could travel "Warp 7" (which geeks will immediately recognize as 656 times the speed of light), it would still take more than 2 days to get to the nearest star beyond our sun, and 152 years to get across our galaxy.

> "It's hard to imagine a more mind-stretching experience than learning, as we have over the last century, that the reality we experience is but a glimmer of the reality that is."
> —Brian Greene, physicist

The speed of light also causes some fascinatingly weird challenges, such as: If you're driving your 1972 Camaro at 99 percent of the speed of light and you flip on the headlights, what happens? The answer, in short, is that you see the light travel away from you at—you guessed it—the speed of light. This conundrum is due in part to the fact that the faster you travel, the slower you become—that is, time slows down for you relative to everything else. Plus, as you speed up, you actually get squeezed in length—you become shorter along the direction you're traveling, called the Lorentz contraction. These are infinitesimal changes at any speed we can actually imagine: Even if you were traveling at 42 million meters per second (95 million mph, or about a seventh of the speed of light), your length would contract by only 1 percent. But accelerate to just below the speed of light, and you would appear from the outside as though you were squished down to a speck's width—and while time would progress normally from your perspective, it would appear as though it had virtually stopped for you from the outside.

Again, these bizarre effects are meaningless at our human scale, but they absolutely must be taken into account in the realm of the atom, where particles and energy fields move and change unbelievably fast. In half a femtosecond—or 500 attoseconds—light travels 150 nanometers, about the size of a virus. A single attosecond is a billionth of a billionth of a second, so there are almost twice as many attoseconds in each second than there have been seconds since the big bang. But it still takes a full attosecond for light to traverse the length of three hydrogen atoms.

Some scientists have proposed that the shortest possible duration—the smallest slice of time that makes any sense or that we would ever need consider—should be called the chronon and measured as the time it takes for light to travel the diameter of a single atomic proton: about 6 yoctoseconds, or six millionths of an attosecond. Proponents argue that every event from the nuclear to the cosmic can be broken down into a series of chronon-long

The special theory of relativity states that an object would require an infinite amount of energy in order to accelerate to the speed of light. But it doesn't actually preclude things that are already faster than the speed of light. Some theorists believe subatomic particles called tachyons may exist that can travel *only* at superluminal speeds, not slower.

segments. It's a compelling idea, but unfortunately quantum physics has opened a world far smaller than the proton. How can we explain the strange workings of quarks and bosons if we're limited by chronons?

Which brings us to the Planck, the smallest possible measurement—anything smaller and our physics truly dissolves in a foam of random probability. The Planck length is a hundred billion billion times smaller than a proton: roughly 1.616×10^{-35} meters. A photon traveling at the speed of light would take one unit of Planck time to cross one Planck length: about 5×10^{-44} seconds—truly the ultimate "quantum of time."

The Trouble with Time Now that we have constructed a carefully calibrated measuring stick for time, with clear marks to help us understand and name any duration or speed, let's apply it to a simple question: How long is the current moment? Perhaps a moment is experienced at our human scale, near a tenth of a second? Or are moments chronon or Planck sized? Or perhaps "now" is longer than we think, lasting an eon?

The trouble, you see, is that we can discuss time all day, but no one—from the scientist to the spiritualist—really has any idea what he's talking about. Every inquiry into time ultimately distills down to a series of questions, like dregs at the bottom of a bottle. For example, here's another one: Is this minute as long as one we had yesterday? While you consider that, remember, too, that our only tool to measure time unfortunately uses circular reasoning: "Time passes at exactly one second per second." Worse, time has a way of slipping by in such a way that we cannot hold one duration up to another to compare them, like we could with two objects. And worst of all, while common sense tells us that time marches forward at the same rate everywhere, physicists now insist that's almost certainly not true. Instead, due to the force of gravity, unless you're lying down, your head is literally aging faster than your toes.

You can blame Einstein for pointing out this odd disparity. In classical physics, Isaac Newton's commonsense science of apples dropping and pendulums swinging, time is an absolute, fundamental structure, like a frame on which we can drape tautly drawn sheets of space. It's a comforting view of time, and one that usually works exceedingly well. But Einstein's theory of relativity points out that how time appears to us from our human perspective is not at all how time behaves everywhere. It's like someone saying, "One plus one equals two in most cases, but sometimes it's a bit more or a bit less."

Relativity tells us that time, intricately interwoven with space, is a flexible medium and that it stretches based on gravity and speed. This implies that two objects moving at two different speeds would travel through time at two different rates, which seems crazy . . . but it turns out to be true.

When the U.S government shot the first global positioning satellite (GPS) into space in the 1970s, no one knew for sure if Einstein's theories were right, but millions of dollars were at stake: If the onboard clock became inaccurate by more than about 25 nanoseconds, it would be effectively useless. However, relativity predicts that a clock traveling through space at the speed of an orbiting satellite should tick far slower than one on the ground—about 7 microseconds per day slower! That alone would cause enough of a problem, but Einstein also predicted that a clock at that altitude—farther from Earth's space-time-warping gravity—would speed up by 45 microseconds each day. The two relativistic effects combined would cause an error of 38,000 nanoseconds each day, enough that a GPS unit on the ground would provide an incorrect reading after just two minutes.

Fortunately, like good investors, the scientists hedged their bets, including a switch on the satellite that could be remotely thrown to enable or disable corrections for relativity. It didn't take long to realize that relativity worked exactly as predicted decades earlier: Speed and gravity affect time.

A clock at the top of Mt. Everest will pull ahead of one at sea level by about 30 microseconds per year.

In today's world of hyperaccurate measurements, scientists have found the same effects here on Earth. They can detect the slowing of time you experience when you ride a bicycle, or how you age faster when you climb a ladder. From our human scale, you never notice the effects, which add up to only billionths of a second over an entire lifetime. But the adjustments are significant when dealing with the ultrafast: Physicists working with atomic clocks must now correct for relativity even when they compare clocks on different floors of the same building. It's not that one clock is more accurate than the other; each clock, just like each of us, literally steps through time differently.

With that in mind, our understanding of "now" takes on an even stranger twist: Just as "here" means "where I am," "now" is intimately tied to "when I am"—time is always personal.

Consider, too, that our dependence on light leads to the realization that there is always a time delay between when something happens and when we can learn about it. If the sun exploded right now, we wouldn't—literally couldn't, due to the speed of light—know for about eight minutes. In other words, relativity says one object cannot affect another object without the passage of time. Add to this inevitable delay the fact that two events that may appear to happen at the same time from one point of reference will appear to be happening at different times from someplace else, and the whole idea of us ever truly grasping a "now" basically falls apart.

As gravity increases, time inevitably slows. Therefore, inside the infinite gravitational field of a black hole, from which even light cannot escape, time stops.

There's No Time Like the Present The dissonance and anxiety every sane person feels when confronted with these facts about time are normal; after all, time is fundamental to everything we do, everything we feel we are. And each of us has a choice: We can simply accept our intuitive sense of time and enjoy what little we have left, or we can stew, dissatisfied with the easy answers, longing for a clearer picture.

On the one hand, Eastern philosophy directs our attention to the present moment—that infinitely small and fleeting slice—as all we have. This is wonderfully encapsulated in the Hallmarkian poem "Yesterday is history, tomorrow is a mystery, but today is a gift—that's why they call it the Present." On the other hand, perhaps the scientist's credo was best summed up by the comedian George Carlin: "There's no present. There's only the immediate future and the recent past."

The truth, if we may use such a bold word, is almost certainly weirder, delving far into a world more commonly explored by philosophers and daydreamers.

Case in point: Physicists have thrown yet another wrench into the time machine. Whereas classical physics describes an absolute clock, and relativity proves there isn't one, quantum physics appears to point, once again, to some larger, external stopwatch synchronizing events throughout the universe. Of course, like everything involving quantum physics, its argument seems to defy logic: A measurement performed on one subatomic particle (such as a photon or an electron) appears to be able to affect another particle elsewhere, simultaneously—as if the second one knew what was happening to the first, ignoring apparently inconsequential things like space or the speed limit.

But what really bothers physicists is that when you lay out all the equations that appear to describe our universe, none of them specifies a "now," or even that there is a future distinct from the past. For a physicist, time does not pass, or flow, or fly—it just is: Past, present, future are all one thing, like a finished "timescape" on a canvas.

In this model—called eternalism—our limited consciousness keeps us from seeing the bigger picture, constraining us to a single moment in time. We hold these moments as special, because they're all we have, all we're able to sense. But eternalism also implies that we're locked to a predetermined future, like actors in a movie playing our parts toward the inevitable (but as yet unknown) conclusion. If this were true, and we were to believe it, would our human sense of urgency and curiosity, our drive, disappear, knowing

A month before his death, Albert Einstein wrote of the recent death of his longtime friend Michele Besso: "Now he has departed from this strange world a little ahead of me. That means nothing. People like us, who believe in physics, know that the distinction between past, present, and future is only a stubbornly persistent illusion."

that none of it really mattered after all? And if so, would that reaction simply have been predetermined, too?

Even a hardened scientist finds this complete lack of free will distasteful. So try this on for size: What if there were multiple universes—in fact a near-infinite number of universes being created every moment—allowing for every possible future outcome to every choice you make. In this model, called the Many-Worlds Interpretation, all time—every event from the Big Bang to the End of It All—remains fixed within each universe like a solid block. And instead of us moving forward through time in one universe, we each move seamlessly from universe to universe, without knowing it, providing an illusion that we're stepping through time. While this no doubt sounds like a bad plot of some science-fiction movie, it happens to be the leading argument among the top minds in academia.

However, one of the hardest problems to solve in any of these eternalist models is why we remember the past but not the future—that is, why we move forward along the arrow of time. If time is just another dimension, like length or height, then shouldn't we be able to traverse it any which way we please?

You may recall from school that the second law of thermodynamics insists that, as time progresses, things get more and more chaotic (entropy increases). So if you spill your milk, it will likely cause a mess. That does seem to imply that events always proceed from past to future. But bizarrely, that's the case only from our macroscopic perspective. If you watched the scene from the atomic level, every interaction along the way, every atom or molecule that jiggles or wags as the milk spreads, could be played forward or backward. After all, it's just as likely that a single molecule will move up or down, left or right, which means that it is technically possible that the milk could flow back into the glass, effectively stepping backward in time. It's only when you look at the mass of the liquid as a whole—the average tendency of trillions of atoms—that flowing "backward" becomes very, very improbable. It's so improbable,

> "He asked me if I knew what time it was. I said, 'Yes, but not right now.'"
>
> —Steven Wright, comedian

Reputable scientists believe that time travel is theoretically possible, by using wormholes in space, infinitely long rotating cylinders, and other clever tricks. However, to generate these kinds of conditions at anything larger than the subatomic level, we would likely have to generate as much energy as an exploding star.

in fact, that we gain this overwhelming sense of time inexorably marching forward.

Nevertheless, at the smallest scales, our conventional sense of time starts to break down completely. Even the concept of the absolute, unchangeable past ("What's done is done!") is beginning to fray. It's becoming clear that, at the quantum level, decisions made today appear to affect events in the past. As the physicists Stephen Hawking and Leonard Mlodinow point out, "The (unobserved) past, like the future, is indefinite and exists only as a spectrum of possibilities . . . The universe doesn't have just a single history, but every possible history, each with its own probability; and our observations of its current state affect its past and determine the different histories of the universe."

With this in mind, it's entirely possible that the future is affecting the past and present all the time without our knowing it, as subtle changes at the quantum scale aggregate to influence our macrocosmic world. For example, in 2009, when the Large Hadron Collider in Switzerland was beginning its groundbreaking work to uncover the hypothetical subatomic Higgs boson, it experienced a major failure and had to be shut down. At the time, "a pair of otherwise distinguished physicists," as the *New York Times* called them, suggested that this may have been caused by the Higgs boson itself, traveling back in time to stop the collider before it uncovered it, "like a time traveler who goes back in time to kill his grandfather." While this may seem absurd, it could explain why everything at the quantum level is necessarily blurry to us; the future needs plenty of wiggle room in order to pull off changes in the present without getting caught.

Ultimately, it's likely that everything we think we know of time— our memories, our carefully constructed calendars, our most precise measurements—is an illusion, a mirage generated like a hologram, to enable us to make sense of a universe stranger than anything dreamt of in any of these philosophies. As Einstein wrote, "The only reason for time is so that everything doesn't happen at once." It may

> **"Whether I come to my own to-day or in ten thousand or ten million years, I can cheerfully take it now, or with equal cheerfulness, I can wait."**
>
> —Walt Whitman, *Song of Myself*

Duration*

5.4×10^{-44} s	Planck time (shortest possible time)
0.3 ys	Lifetime of W and Z bosons
6 ys	Light travels diameter of atomic proton (1 chronon)
1 as (1 million ys)	Light travels length of 3 hydrogen atoms
12 as	Shortest laboratory laser pulse on record
320 as	Electron transfers from one atom to another in an electrical reaction
1.3 fs	One cycle of electromagnetic light between visible and ultraviolet light
200 fs	Fast chemical reactions (such as eye reacting to light)
1 ps (1 million as)	Half-life of a bottom quark
3.3 ps	Light travels 1 millimeter
1 ns	One machine cycle on 1 Ghz computer chip
2.5 ns	One million wavelengths of red light
5.4 µs	Light travels 1 mile in vacuum
22.7 µs	Length of a sound sample on an audio CD
5 ms	Bee wing beats once
8 ms	Camera shutter speed at 1/125 s
33.3 ms	One frame in digital movie
41.7 ms	One frame in a film movie
200 ms	Average human reflexes
30 cs	Blink of an eye
43 cs	Fastball travels from pitcher's hand to home plate
1 second	Human heartbeat; light travels 300,000 km
9.58 s	World record 100 m dash
10.5 minutes	About the half-life of a neutron outside an atom
1,039 seconds (17 minutes, 19 seconds)	Record time holding breath underwater
28.8 ks (8 hours)	Average human daily sleep requirement
86.4 ks	One day
29.5306 days (2.55 Ms)	Lunar month
40 days	About the longest a person can survive without food
125 days	Life of red blood corpuscle
23 Ms (38 weeks)	Length of human pregnancy
356.2422 days	Average solar ("tropical") year
27.7 years	Half-life of strontium 90
75 years (2.3 Gs)	Typical life span for a human being
90 years	Life span of anemone (longest-living invertebrate)
3.16 Gs	One century

122 years, 164 days	**Oldest human: French woman Jeanne Calment (1875–1997); that's 3.86 billion seconds!**
150 years	**Life span of tortoise**
164.8 years	**Neptune's solar orbit**
248.09 years	**Orbit of Pluto around sun**
550 years	**Time since humans learned to roast coffee**
31.55 Gs	**One millennium**
2,540 years	**Time since Buddhism founded**
6,000 years	**Time since humans learned to brew beer**
11,800 years	**Time since last ice age (Holocene epoch)**
25,784 years	**Earth's axis returns to same location (precession of the equinoxes)**
1 Ts (10^{12} s)	**31,689 years**
100,000 years	**Time since Homo sapiens appeared**
65 million years	**Cenozoic period (time since the dinosaurs died off)**
225 million years	**One rotation of the Milky Way**
710 million years	**Half-life of uranium 235 (relatively rare)**
1.26 billion years	**Half-life of potassium 40 (we have billions of these in our bodies)**
2.4 billion years	**Time since the Great Oxygen Catastrophe**
4.5 billion years	**Age of Earth**
4.51 billion years	**Half-life of uranium 238**
4.63 billion years	**Age of sun**
13.75 billion years (about 434 Ps)	**Age of universe**
10 trillion years (312×10^{18} seconds)	**Estimated life span of a red dwarf star**
100 trillion years	**Stelliferous era (time until all stars burn out or collapse into black holes)**
311.04 trillion years (about 9.8 Zs)	**Lifetime of Brahma in Hindu mythology**
10^{34} years	**Estimated half-life of proton**
2×10^{67} years	**Life span of a small black hole (mass of the sun)**
10^{100} years	**Time until stars, galaxies, black holes, and virtually all matter in the universe cease to exist**

*You can refer to the table of prefixes and their abbreviations on the last page of this book.

even be that the mind itself creates and manages the flow of time. We don't know, and may not even be able to know, the answer.

From Here to Eternity We live at a rate that allows us to experience the imperfect metronome of daily tides, of seasons changing, of trees growing. But we cannot watch the blossoming of a flower or follow a bullet's trajectory with our own eyes—those speeds are outside our spectrum of ability. Nevertheless, the worlds of the very fast and the very slow are just as real as ours. Does a turtle think it's walking slowly, or a hummingbird have any idea how quickly it's moving? Or does the hummingbird watch us in wonder at our snail's pace?

The neurologist Oliver Sacks has written movingly of a patient in a near-catatonic state who sat for hours in one position. Only much later did Sacks realize that the man was moving: He was wiping his nose, but at a rate that took an hour to bring his hand to his face. Another patient lived on the opposite end of the spectrum, having no trouble catching a fly in midair because it appeared to her that the insect was moving lazily. In neither case did the patient think their time was any different from ours. Today, doctors are starting to find that time—or, more specifically, a mismatch in our personal sense of time and the time moving around us—may be at the heart of a wide range of physical and psychological disorders, including Parkinson's disease, attention-deficit/hyperactivity disorder, and even some forms of autism and schizophrenia.

Time is, without a doubt, a crucial element of our consciousness and our ability to understand not just the world around us but also the extraordinary worlds we can track with instruments. After all, only through time can we experience change, and only through the aggregate of the tiniest and most subtle changes can the magisterial movement of galaxies be achieved.

Of course, time isn't reserved for just the mundane workings and measurements of atoms or eons. The twentieth-century Jewish theologian Abraham Joshua Heschel wrote that God was to be found not in sacred places but rather in sacred time. If we're searching for meaning, or for a greater connection to the universe in which we live, it's clear that we must move beyond a comfortable and passive acceptance of time's mysteries.

What if 13.7 billion years, or even a trillion years, is but a single lifetime, one of many thousands? Perhaps the age of our universe is the equivalent of a single Planck time in another, greater universe, and the cosmos we know, the billions of stars living and dying, is but a momentary blip into existence before it snuffs out, all in a moment too small to even measure. Does that mean our human scale, our human time, is any less important, any less glorious? It's all a matter of perspective.

EPILOGUE

Imagination is more important than knowledge.

–Albert Einstein

WE ARE THE ONLY SPECIES WE KNOW OF THAT CAN IMAGINE, SO imagine with me: Imagine being a neutrino that, while traveling through even the most dense rock, is in wonder at the vast spaces within and between atoms, like a spaceship traveling between the planets and the stars. Imagine a galaxy enjoying the delicious sensation of its several billion star parts twisting and turning, like an early morning stretch. Imagine a raucous midair party game, as one molecule passes on messages by playfully nudging its neighbors, propagating the sound wave that is about to tickle your eardrum.

Imagine you can watch two atoms as they exchange electrons in a slow dance of attraction, then turn to see the swirling flow of mountains rising and eroding through the swiftly shifting power of water and air across an eon of time. It's all true, it's all happening right now, though we can experience these visions only through our mind's eye.

And now, look around you. Like so many real-world adventures, one of the best parts of embarking on a journey into outer and inner space is the return home, where we can rest into our comfortable human-scale senses, enlightened by our new awareness that the world is far more complex, fun, and—yes—intimidating than we had thought. But now returned, how are we to understand our place in this vast spectrum?

We see a rose and attempt to understand its roseness by breaking it down—cell by cell, molecule by molecule, atom by atom—to its

> "Science cannot solve the ultimate mystery of nature. And it is because, in the last analysis, we ourselves are part of the mystery we are trying to solve."
>
> —Max Planck, physicist

essence, and we discover that it is miraculously constructed from star dust, remnants of supernovae billions of years ago. And not only is it the same stuff we're made of, we find that at a certain level it's difficult to determine where *it* stops and *we* begin.

But as we draw our vision back, enhanced by our newfound insights, we find an even more astonishing perspective: The elegance of a single rose is different from the beauty and magnificence of a rose garden of ten thousand bushes. Neither is more beautiful, neither is more real; both must be taken as part of the whole and yet wonderfully separate.

It's like exploring a single note played within a Bach fugue versus its place in the theme, versus the piece as a whole, versus the canon of Bach's works, versus the whole of Baroque music. Or, conversely, considering the attack of that single note, its duration, its tremolo, its resonance. Can the entire splendor of the Baroque period be captured in the resonance of that single note? Can the amazing clarity and richness of the single note be adequately reflected in the waves of millions of notes from that period? Or is each perspective, each range, as valuable, as rich, as sublime?

So it is in our lives, caught here in the middle world between large and small, cold and hot, slow and fast. We seem insignificant—less than a speck on a speck, and still hardly able to see, much less manage, the world of the atom or that of the universe. But appearances can deceive. Is a star any more important than we are because of its great size or power? A loud sound can demand our attention, but does it necessarily convey more than a soft one? We humans have something neither the cosmos nor the atom has: mind. We can appraise and appreciate; we are curious and cunning; we delve and dream. And who is to say that in the grand scheme of things—whatever collection of dimensions and multiverses we truly live in—these qualities of mind are not just as meaningful as mass or momentum?

You might even say that creativity and communication are our force carriers, the way photons act within and between atoms, or

> "I regard consciousness as fundamental. I regard matter as derivative from consciousness. We cannot get behind consciousness. Everything that we talk about, everything that we regard as existing, postulates consciousness."
>
> —Max Planck, physicist

gravity plays between the stars. It is only from our reference point in the middle that we are able to survey and consider a panorama of spectrums.

Unfortunately, it's also true that our minds may not be ready to grasp many of the clues we're uncovering. In fact, most of what we have learned about the universe is literally inconceivable—our brains do the math, we can logically accept and argue it, but it's too weird, too astonishing to truly grok. So on the one hand, there is no doubt that we know far more than people a century or two ago; but on the other hand, we find ourselves still somewhere in the middle of a spectrum of understanding, or even of sentience.

We so want to believe that if we understand the parts, we will understand the whole—and if we understand the whole, we'll finally understand our place within it. If only we could gather enough data, we'd be able to forecast the weather, or understand time's arrow, or explain the delicious laughter of a nine-month-old playing peekaboo. It's a powerfully compelling concept, but twenty-first-century science is waking up to the idea that it's not going to happen anytime soon—or perhaps ever.

We're at the tipping point where much of what we think of as "truth" is being overturned, and our children are growing up in a world of "what-ifs" and "we cannot knows." To manage, we must learn to be curious adventurers, letting our left brain crunch the numbers while developing our right brain's ability to poetically intuit these extraordinary realms, if not to find answers then at least to find meaning.

If we've learned anything, it's that beyond every horizon is another mountain, inviting us to learn more and delighting us with new vistas. We must simultaneously surrender into the mystery, celebrate our achievements, and strive for more. For that's what we humans do as we compare ourselves to others, making spectrums of our experience that inspire with an almost spiritual sense of wonder.

> "Penetrating so many secrets, we cease to believe in the unknowable. But there it sits nevertheless, calmly licking its chops."
> —H. L. Mencken, satirist

ACKNOWLEDGMENTS

> "Even though you can't see or hear them at all, a person's a person, no matter how small."
>
> —Dr. Seuss, *Horton Hears a Who!*

I AM DEEPLY GRATEFUL TO A WIDE SPECTRUM OF INDIVIDUALS and organizations for their inspiration, education, and cooperation. First of all, my thanks to my father who art in Austin, Adam Blatner, whose brainstorming and encouragement were invaluable. My agent, Reid Boates, who talked me into what became the hardest project of my career, and publisher, George Gibson, who believed I could pull it off. And to my editor, Lea Beresford, and designer (and longtime friend), Scott Citron, who were crucial in the process. Thanks, too, to Nancy Chamberlain, Cynthia Merman, and Nicole Lanctot for checking my work and cleaning this up, and to Lisa Silverman for shepherding the process beautifully.

Many thanks to the American Museum of Natural History, the King County Library System, and Amazon.com, who fed my habit. A shout out of thanks to the Lyons' Den, Peets Coffee, and Café Ladro, without whose "inspiration juice" this book would have had far fewer adjectives; to Daft Punk, Delerium, Garmana, and Jean Michel Jarre, who gave it a beat; to the makers of DEVONthink Pro, who helped me keep all the pieces together; and to the beautiful Inn at Langley, where the final words were typed. To my friends and family, including Gabriel, Daniel, Mom, Richard, Allee, Don, Snookie, Suzanne, Damian, Lucia, Alisa, Paul, Camille, Zoe, Edna, Ted, Ruth, Glenn, Jeff, Mark, and Anne-Marie.

And my deepest appreciation to my wife and partner, Debbie Carlson, who reminds me that words are important and life is magical.

ABOUT THE AUTHOR

DAVID BLATNER IS KNOWN WORLDWIDE FOR HIS AWARD- winning books, including *The Joy of Pi* and *The Flying Book*, and his lectures on electronic publishing. More than a half-million copies of his books are in print in fourteen languages. He and his wife and two sons live outside Seattle, Washington.

FOR FURTHER READING

For additional links, facts, and information, visit: www.spectrums.com

Selected Books

Asimov, Isaac. *The Measure of the Universe: Our Foremost Science Writer Looks at the World Large and Small.* New York: Harper and Row, 1983.

Carroll, Sean. *From Eternity to Here: The Quest for the Ultimate Theory of Time.* New York: Dutton, 2010.

Davies, Paul. *The Goldilocks Enigma: Why Is the Universe Just Right for Life?* New York: Mariner Books, 2008.

Greene, Brian. *The Fabric of the Cosmos: Space, Time, and the Texture of Reality.* New York: Vintage, 2005.

Joseph, Christopher. *A Measure of Everything: An Illustrated Guide to the Science of Measurement.* Ontario: Firefly Books, 2006.

Kaku, Michio. *Hyperspace: A Scientific Odyssey Through Parallel Universes, Time Warps, and the 10th Dimension.* New York: Anchor, 1995.

Potter, Christopher. *You Are Here: A Portable History of the Universe.* New York: Harper Perennial, 2010.

Robinson, Andrew. *The Story of Measurement.* New York: Thames & Hudson, 2007.

Streever, Bill. Cold: *Adventures in the World's Frozen Places.* New York: Back Bay, 2010.

INDEX

Prefix		Size	Name
yotta-	Y	10^{24}	septillion
zetta-	Z	10^{21}	sextillion
exa-	E	10^{18}	quintillion
peta-	P	10^{15}	quadrillion
tera-	T	10^{12}	trillion
giga-	G	10^{9}	billion
mega-	M	10^{6}	million
kilo-	k	10^{3}	thousand
hecto-	h	10^{2}	hundred
deca-	da	10^{1}	ten
deci-	d	10^{-1}	tenth
centi-	c	10^{-2}	hundredth
milli-	m	10^{-3}	thousandth
micro-	μ	10^{-6}	millionth
nano-	n	10^{-9}	billionth
pico-	p	10^{-12}	trillionth
femto-	f	10^{-15}	quadrillionth
atto-	a	10^{-18}	quintillionth
zepto-	z	10^{-21}	sextillionth)
yocto-	y	10^{-24}	septillionth